Environmental Participation

Catharina Landström

Environmental Participation

Practices engaging the public with science
and governance

Catharina Landström
Department TME
Chalmers University of Technology
Gothenburg, Sweden

ISBN 978-3-030-33042-2   ISBN 978-3-030-33043-9   (eBook)
https://doi.org/10.1007/978-3-030-33043-9

This Palgrave Pivot imprint is published by the registered company Springer Nature Switzerland AG
The registered company address is: Gewerbestrasse 11, 6330 Cham, Switzerland

Acknowledgements

This book has been made possible because many people have been willing to get involved with environmental participation. I thank my colleagues in the university whose guidance, advice and support continue to keep transdisciplinary environmental research exciting. I owe even more to all the local people who have generously contributed time and effort participating in the research projects that have provided me with the practical experience upon which the book relies.

CONTENTS

CHAPTER 1

Introduction: Environmental Participation

Abstract I present the science studies perspective used in this book and explain why environmental participation is addressed as a distinct type of public engagement. Origin stories of environmental participation are outlined and a practice-based distinction between environmental participation in science, decision making and expertise is introduced. This distinction is explained and applied in a presentation of the key concepts used to discuss participatory practices in this book. The characteristics of participating publics are considered and finally, I outline the content of the subsequent four chapters.

Keywords Environmental participation · Science studies · Publics

A SCIENCE STUDIES PERSPECTIVE ON ENVIRONMENTAL PARTICIPATION

This is a book about some of the ways in which people engage with the natural environment, acting collectively in and through the institutions of science and democratic government. It is not about individual experiences of the natural environment, or cultural values. It is about how people not formally connected with science and government can get involved with the ways these institutions act in relation to the natural environment. Environmental participation is a notion intended to capture the diverse ways in

© The Author(s) 2020
C. Landström, *Environmental Participation*,
https://doi.org/10.1007/978-3-030-33043-9_1

1

which publics get involved with the work done in institutions to govern the society–environment relationship. The four social constituencies discussed in this book—scientists, decision makers, experts and publics—all engage with the natural environment, in their own ways. Scientists, decision makers and experts involve in professional practices and environmental participation occurs when publics become involved with the activities of these professionals.

Coming to this phenomenon as a science and technology studies (STS) researcher my perspective on environmental participation originates in transdisciplinary research projects, involving collaborations of natural and social scientists and local publics. As an STS researcher, my work is firmly anchored in qualitative social science and my interest is in the many different ways in which we create knowledge about the natural environment. Working together with other social scientists, natural scientists and local people have shown me that we really can do more and better together. Collaborating with people from different backgrounds have forced a rethink of the way I do STS. In collaborative environmental projects it is not possible to study the creation of scientific knowledge and its role in society from the outside. All project members have to contribute to the shared goals. Doing STS differently in projects with environmental participation has provided the motivation for this book that focuses on practical experience, not societal dynamics.

Working with environmental issues in collaborative projects has also alerted me to the importance of place. Environmental experiences are always geographically specific, there is no general environment outside of the pages of academic journals and books. In reality it is always a particular environment, with its unique configurations of land, water, air, plants, temperature, animals, humidity and so on. STS traditionally focus on science that strives to generate general knowledge and technology that can be applied anywhere has not paid much attention to the specificity of place. Fortunately, I could learn from human geographers in transdisciplinary projects and in this book STS perspectives are supported by concepts from human geography. The hybrid environmental STS perspective resulting from this combination is the starting point for this book that attempts to tell a coherent story about environmental participation in practice.

In this introductory chapter I start by explaining why it is useful to consider environmental participation as a distinct area of public participation. After this I retell some of the origin stories of the field that are told by STS researchers. The existence of a multitude of such origin stories points to the fact that environmental participation is a novel and growing topic in current academic debates. However, this is not a book that aims to make an impact in these debates, hence, they will only be dipped into to offer some concepts that are useful to organise the presentations of environmental participation practices which are my main interest. After outlining STS concepts capturing environmental participation in science, decision making and expertise we will look at the discussion about participants. And, finally, the four subsequent chapters of the book are outlined.

Why Environmental Participation Is Unique

The previous section hinted at the reasons for viewing environmental participation as a distinct area of public engagement. This is a rather unusual view, mostly public participation is treated as the same, regardless of what people participate in, when discussed by social scientists. This makes sense when the focus is on the dynamics of power in participatory processes, however, there are important reasons for distinguishing between public participation in different fields when the interest is in the practices.

One of the features of environmental participation that stands out to me is that place, in a geographical sense, matters. The environmental processes and problems that become the subject of scientific research, decision making, expert management and public concern occur in particular locations. In contrast many STS cases studies of public participation have highlighted medical research that concerns patients with a particular condition that does not relate to geographical location. With regard to environmental issues the features of the location are important determinants for the nature of the problem, for example, flooding is experienced by some people, but not others, and flood risk matters in places close to rivers, lakes or the coast (although flash flooding due to heavy rainfall can occur anywhere). In these places institutional actors will address flood risk and local publics will be interested in how the issue is managed. The ways in which flood risk is governed will depend on the physical features of the flood process, as well as the economic and technical resources available. In places at risk for flooding or where floods have occurred publics will be interested in participating in research and decision making addressing it. Public interest in a local environmental matter of concern to them does not translate into a general interest in environmental science or decision making. Institutions pursuing environmental participation are forced to pay attention to local matters of concern.

The independent agency of the natural environment has to be recognised in environmental participation. Many STS discussions of public participation address policy choices, for example, about how to regulate new technologies. In contrast environmental participation concerns processes occurring in nature that are not fully knowable or controllable by science or decision makers. Environmental processes are complex and emergent, scientific knowledge about them is tentative and evolving. Environmental governance is never complete, as the natural environment continues to change new management strategies and interventions must be developed in response. Environmental participation often occurs in contexts where the non-human independent physical agency of environmental processes pose risks to humans, such as flooding. These risks may originate in environmental change caused by human activity, but they must be analysed as non-human agency and addressed with some type of physical intervention. The matters of concern in environmental participation are always more-than-human, involving non-human agency outside of institutional control.

The role of the natural environment means that environmental participation always involves at least three parties—processes in nature, institutions and publics. The relationship between the first two is constitutive, it is in the institutional activities engaging with the natural environment that publics can participate. The notion of participation assumes that there is organised, goal-directed activity with which publics can get involved. When bringing lay people into the relationship of institutions and the natural environment it becomes clear that everybody experiences the physical environment. The publics becoming involved in environmental participation have had time to develop experiences and knowledge by engaging with environmental processes in a place they are familiar with. This is very different from publics invited to participate in deliberations on how to govern or use, e.g. nanotechnology, they have no experience of encountering these technologies in everyday life before taking part in scientific engagement events.

I do not claim that environmental participation is unique because it is different from participation in other fields, but that it brings together many of the features of participation in a particular way. That publics have independent and sometimes superior knowledge to scientists can also be the case in medical research, patients can know more about how an illness affects the body than medical researchers. The agency of the subject at hand has been an issue in public participation addressing genetically modified organisms. However, in the STS literature public participation has largely been treated as the same regardless of the matter of concern but when the interest is focussed on practices it is beneficial to distinguish between different topic areas.

Environmental participation brings together people who have separate, different relationships with the environment, such as local people affected by a problem, businesses managing potential risks, scientists investigating environmental change, policy makers balancing economic interests against environmental protection and so on. Everybody brings knowledge emerging in their own relationship with the environmental processes to the participatory process.

Origin Stories

While historians would undertake a proper empirical investigation to analyse the background of environmental participation today my understanding of its origins relies on accounts provided by STS researchers and others discussing public engagement with environmental science and governance. Trusting my STS colleagues to strive for accuracy I am fully aware that these are very particular stories, crafted with the emphasis on providing backdrops for analyses of present-day scientific knowledge creation and environmental governance. And, as expected these authors do not talk about 'environmental participation', a term introduced in this book, but about public participation in science and decision making in Western democracies, with an empirical focus on environmental issues.

Introducing their well-received and widely read anthology Jason Chilvers and Matthew Kearnes (2016) understand public participation as the expression of 'a more reflexive pattern of relations between science and the projects of political ordering' (Chilvers and Kearnes 2016: 1). This reflexivity is seen as prompted by, on the one hand, the failure of science and engineering to solve the complex problems facing society that they had been complicit in causing, such as chemical pollution. On the other hand, it has become clear that this failure of science and engineering also troubles political systems and procedures that are rendered unable to protect their citizens against environmental hazards. Starting in the 1960s this problematic has become ever more obvious and today climate change is recognised as the most urgent, yet seemingly politically unsolvable, environmental problem. In response to the failure of the institutions to solve problems 'science *and* democracy have been increasingly opened up to diverse forms of public engagement and participation and wider civic scrutiny' (Chilvers and Kearnes 2016: 2).

Chilvers and Kearnes tell an origin story with focus on the changes in the ability of institutions to effectively and adequately address the challenges facing society as drivers. In contrast, Mathieu Quet (2014) brings the social movements of the 1960s and 1970s to the forefront. Quet traces the roots of participatory discourse to the radical science movement that emerged in connection with the anti-nuclear protests, anti-colonialist movements and protests against the Vietnam War in the affluent post-war Western societies. In a case study of France, Quet finds that activists were wary of the term 'participation' that was perceived as belonging to a discourse of co-optation of workers into capitalist production goals. However, the radical science movement in the 1970s demanded not only 'science "for the people", but also science "by the people"' (Quet 2014: 633). Quet argues that there has been a de-radicalisation of participation since the 1970s to the present day understanding of it as 'deliberation, precaution and risk assessment' (Quet 2014: 640).

Further STS origin stories bring the tension between science and democracy into view. The starting point is the concern about decreasing public trust in science that spread among scientific institutions in the 1990s. Around the turn of the twenty-first century public participation was recast as a way to improve the appreciation of science in society. Scientific institutions in Europe and the United States perceived the public as distrusting and critical of science, they responded by promoting public understanding of science, public engagement with science and public participation as ways to increase the wider appreciation of science. This origin story plays an important role in the critiques of public participation articulated in the early 2000s. Brian Wynne (2006), for example, argues that these efforts did not involve an intent to change the conduct of institutions which was what caused the public to lose trust. In his view the true objective of the public engagement activities at the time was to manipulate the public.

In the STS discussion environmental issues are frequently used to illustrate aspects on discussions of public participation, in contrast human geographer Sally Eden's (1996) origin story focuses on the environment. Her analysis addresses the ways in which environmental movements forced politicians to put environmental protection on the political agenda and made public participation an important aspect of Western democracy. At the 1992 UN conference on Environment and Development in Rio public involvement was explicitly identified as a necessary part in environmental policy and its implementation. As Eden puts it: 'Successful environmental policy has therefore been linked to the notion "concerned citizens", coupling individual action to institutional change in the name of environmental protection' (Eden 1996: 184). Eden's human geography perspective has a wider scope than that of STS researchers and she illuminates the way in which science intersects with environmental participation, as it complicates public participation in environmental policy. She explains how the 'scientization of environmental problems', meaning that they are identified by scientific research and the solutions offered rely on scientific and engineering expertise, has led to a dominance of scientific expertise in environmental policy relegating public participation to the implementation of solutions developed by experts and adopted by politicians.

Topic-specific origin stories of public participation are also told in research on environmental management. Mark Reed (2008) historicises the academic characterisation of different types of participation, beginning with Sherry Arnstein's 'ladder of participation' from 1969. This normative approach to participation classifies activities on a scale according to the degree of control the lay participants exercise over the process.

The overview of origin stories highlights the multiple starting points for environmental participation another important feature is the change over time. With public participation becoming common in environmental management the relationship between science, publics and governance changes. The explicit ambition to include lay publics in the previously quite technocratic, expert-led, environmental management requires new ways of working and changes in spatial and temporal organisation. Key in this change is the re-evaluation of publics, from having been seen as lacking relevant knowledge to being considered able to contribute different, yet important knowledge, to the decision making process. Eden (2017) discusses the changing understanding of environmental publics, explaining how people develop knowledge by engaging with nature in other ways than scientific research. Understanding knowledge relevant to environmental decision making as constituted in everyday activities recognises that 'people may regularly experience a particular environment and build up a store of knowledge and feelings about it' (Eden 2017: 21).

The value of experience-based environmental knowledge, often referred to as 'lay' or 'local', is gaining recognition among scientists and environmental managers. Public participation in environmental science is motivated by the idea that lay people are able to contribute accurate knowledge about a particular local environment that can help scientists generate better scientific knowledge. What shape the contribution of lay people can vary, but citizen science is one format for participation that scientists agree is valuable. Reviewing citizen science in hydrology and water resources Wouter Buytaert et al. (2014) conclude that there is 'large potential for increasing involvement of citizens in data collection' (Buytaert et al. 2014: 18).

Beyond increasing the involvement of lay people in scientific research lies the challenge of participation in the knowledge used for environmental management. This has been a topic for discussion and analysis in the field of environmental management in the last decade. The challenge is not just to involve lay publics but to integrate their experience-based knowledge of local environmental problems with the science-based knowledge in expert assessments. Christopher M. Raymond et al. (2010) probe this issue and argue that it is important to identify different epistemological beliefs regarding what counts as environmental knowledge.

These stories about the origins and different issues comprise a backdrop mosaic for the discussions in this book. It is important to retain a sense of multiple, diverse origins from which stories can be crafted to suit different purposes and avoid the illusion of a singular historical trajectory. Environmental participation is always tied to local conditions, national legal and political dynamics, and to specific geographical materialities. Environmental participation is about local publics getting involved with knowledge creation and governance of the environmental processes and problems occurring in the place where they live. In spite of diverse origins, differences in focuses and formats, there are many similarities in the ways people who are not scientists or policy makers engage with science and decision making to address environmental challenges. These similarities make it possible to discuss environmental participation as a particular field of practice.

Participation in Environmental Science, Decision Making or Expertise?

The notion environmental participation points to a wide range of practices, but the approach in this book is based in an STS perspective that takes a particular interest in science, knowledge and expertise and their roles in society. However, the focus on environmental participation prompts a tweak to the STS framework. One of the core claims of STS, as a field, is that science is not separate from other parts of society and that knowledge production must be analysed in its social context. However, on the less abstract level of environmental participation there are important differences between science, decision making and expertise. The distinction I make between these three areas of practice is based on experience. Working in several scientific projects involving participation I have learned that, although participatory research could eventually impact on environmental management, this is not an automatic process. This is not a trait specific to participatory research, many environmental scientists want to believe that the knowledge they create will be welcomed and put to use by decision makers but often find that this is not the case. This is also true for participatory research, in many cases the outcomes are confined to the scientific knowledge produced and never taken up in decision making. In contrast public participation in environmental decision making has a very different relationship to society and is organised through formal procedures and legal frameworks. Impacting on decision making in a democratic society is serious business and it has to be done properly, in ways that do not undermine democratic principles.

The difference between participation in science and decision making has important practical implications. For example, scientists can recruit participants based on criteria relevant for their research project, but participants in decision making need to be representative for the affected population. Approaching the different fields in practice it is possible to distinguish a scientific research project from a management intervention, particularly when scientists fail to have the impact on the management intervention that they envisaged when designing the research project. It is less easy for most of us to see the difference between science and experts, but they do relate very differently to decision makers. Science-based expertise is directed towards the specific, the decision makers requesting expert advice want it in order to address a particular issue that they need to act upon. Experts are commissioned to provide actionable knowledge. Decision makers sometimes hide behind science and expertise to gain authority and legitimacy for political, value-based decisions. However, they have to consider transparency and representation to preserve the democratic process. The difference between these fields is obvious when we engage with local practice rather than historical development or institutional dynamics. Even if science and decision making are closely connected and historically entwined, we can spot the difference in the here and now, i.e. in practice. Since the focus in this book is on environmental participation in practice it makes sense to distinguish between science, decision making and expertise, in this particular context.

Public Participation in Environmental Science—Co-production, Dialogue or Education?

Distinguishing between environmental participation in science and decision making is only the first step. In each of these distinct fields of practice we need to make further clarifications with regard to the type of activity they involve. We start with looking at how publics can participate in science.

A very useful and widely adopted typology was introduced at the end of the 1990s by Michel Callon. Discussing the role of lay people in science he identified three 'models' that he called 'co-production', 'dialogue' and 'education' (Callon 1999). The co-production model was the most radical as it involves scientists and lay participants working together as equals, generating new knowledge together. In chapter two we will discuss co-production practice extensively, interrogating the principles, applications and effects of such experiments in knowledge creation.

Callon's second model 'dialogue', involves scientists and lay participants getting together to talk. The aim is to exchange knowledge, information, interpretations, points of view and opinions. 'Dialogue' is not intended to change the way in which science produces knowledge, but to create spaces in which science and publics can meet to trade and align the things they know. Outcomes of such interaction can be new uses for scientific knowledge and science being packaged in ways that are useful for particular purposes. This form of participation will also be examined in chapter two as we consider examples and reflect on how dialogue practices impact on environmental science.

The third model for scientists' interaction with publics that Callon called 'education' is a lot less clearly defined. Because the context of the conceptualisation is public engagement with science, we assume that it is not a question of education in a formal sense, e.g. schools and universities. Instead it is useful to understand it as referring to the direction of the flow of knowledge in situations where scientists and lay people meet face to face. In the context of public participation the 'education model' can be applied to activities in which the scientists teach people what to do. While Callon talked about science fairs and such activities as educational interaction of scientists and lay publics, environmental participation is about lay people doing something active. Elaborated in this manner the notion of participation as 'education' can capture key aspects of citizen science.

Citizen science is big in environmental science; all over the world lay people take part in many different types of activities, ranging from searching for invasive species, to counting insects, to classifying newly discovered species. In chapter two we look closer at some examples of citizen science in practice and consider its role for scientific knowledge production.

Co-production, dialogue and education are terms developed to capture the role of lay people in scientific research. In chapter two these concepts provide structure to the discussion of the diverse ways in which publics participate in environmental science. In the final, fifth chapter, we return to these concepts to reflect on how focussing on practice changes our understanding of them by making us see what they do not capture.

Public Participation in Environmental Decision Making—Instrumental, Substantive or Normative?

The distinction between environmental participation in science and decision making made in this book allows for clarity regarding the applicability of three widely used concepts in STS discussions about participation—normative, instrumental and substantive rationales. I use them with reference to public participation in environmental decision making. Insisting on this delineation we can draw on science policy expert Andy Stirling's explanation of, to start with, a normative rationale for public participation as resting on principles of 'democratic emancipation, equity, equality and social justice' (Stirling 2007: 220). This is a claim for everybody's democratic right to take part in shaping decisions that affect them.

An explicit normative rationale can be found in national and international environmental policy, for example, international treatises such as the Aarhus declaration, and environmental policies in the European Union, such as the Water Framework Directive. From a normative perspective public participation in environmental decision making is a good in itself and it does not need any further justification. When moving from policy to practice the challenge a normative rationale poses to environmental decision making is to make sure that everybody who is affected can and do participate. Comprehensive participation is, of course, impossible but where should the line for what is reasonable be drawn? How many and which publics are needed for a normative rationale to be fulfilled? It is very difficult to succeed with involving all affected publics and most events organised to ensure environmental participation in decision making can be criticised for not being inclusive enough and failing to involve marginalised people.

In chapter three we look at the West Cumbria MRWS Partnership that undertook a very wide range of activities in their pursuit to engage with everyone in the locality on the issue of geological disposal of radioactive waste. We will also see how this local organisation operated according to a normative logic while the institutional actors that created the conditions for its existence acted on its opposite—an instrumental rationale.

The instrumental rationale is found in participation based on the idea that to involve publics in decision making is a 'better way to achieve certain ends' (Stirling 2007: 220). By involving the public decisions are believed to become more legitimate. In the case of the West Cumbria MRWS Partnership the UK Government was clearly adopting public participation as a way to achieve the objective of geological disposal of radioactive waste. Public opposition had made implementation impossible and when restarting the efforts, participation became an attempt by the Government to mitigate negative reactions and gain legitimacy.

Participation driven by an instrumental rationale aims to further the interests of the decision making institution. The purpose is to create wide societal consensus about policies and decisions to facilitate implementation and confer public legitimacy. However, the complexity of decision making means that what originates in an instrumental rationale may very well be turned into something else on a different level in the democratic system. The West Cumbria MRWS Partnership is one example that will be discussed in chapter three, another is the Catchment Based Approach (CaBA) in water management.

CaBA began as the strategy by which the UK Government tried to implement the normatively underpinned requirement for public participation in the EU Water Framework Directive. It was introduced by the Department of Environment, Food and Rural Affairs (Defra) in the early 2010s, first in a few pilots and subsequently rolled out across the country. However, in 2019 CaBA has become a way for local, community-based, environmental stewardship organisations to have a say on how to best manage local water environments in collaboration with water utility companies, expert advisors and local authorities. In this instance a normative rationale in the EU was turned into an instrumental motive in the UK that became a substantive rationale on catchment level.

The third—substantive—rationale is at hand when participation is promoted because it is thought to lead 'to better ends' (Stirling 2007: 220). A substantive rationale often underpins activities to which publics are invited to contribute ideas for how to solve problems and input to decisions about how to move forward with environmental management issues. The purpose is to achieve better decisions by widening the range of knowledge and values involved. The institutions inviting publics to substantive participation conceive of practical and experience-based knowledge as equally important as scientific and technical expertise in finding solutions to complex real-world problems. Today this attitude permeates the presentation of CaBA on the official website[1] and it seems as if participatory catchment management has become established in many locations across the UK.

Whether public participation in environmental decision making is underpinned by a normative, an instrumental or a substantive rationale makes a significant difference for the participatory practices, but what remains a fact is that the invitation is issued by the formal institutions in power.

Uninvited or Invited Participation in Expert-Led Environmental Management?

The critical distinction made between environmental participation is science and in decision making could risk reviving old notions insisting on science being value-free and autonomous. This is far from my intent, the point is to be able to distinguish in which ways different activities can be expected to bring about change which is critical in relation to environmental participation. Lay people who contribute time and effort deserve to know what difference their participation can make. It really matters if the impact will be on the creation of a scientific modelling perspective reported in a scientific journal, or if it will affect the construction of flood protection measures in a local catchment. Many scientific projects expect the outcome of local research to impact on decision making, but they are most often wrong. Scientific projects very rarely affect environmental decision making directly. Most often scientific knowledge is brought into environmental decision making by experts who apply it to local circumstances. Expert-led environmental management is the third arena for environmental participation. It is also the site at which science and decision making intersect most visibly, hence, participation here is often born of controversy.

Brian Wynne (2007) has introduced a distinction between 'invited' and 'uninvited' participation, founded in a discussion of controversies over environmental risks, often created by corporate action and/or institutional

---

[1] https://catchmentbasedapproach.org/.

inaction. He explains that '[D]eliberately or not, invited public involvement nearly always imposes a frame which already implicitly imposes normative commitments—an implicit politics—as to what is salient and what is not salient, and thus what kinds of knowledge are salient and not salient' (Wynne 2007: 107). Case studies have shown that invited participation usually means that the actors extending the invitation remain in full control of the format, the content and the outcomes. Being able to choose the site, the timing and the agenda of participation contribute to generating the outcomes needed by the institutions.

In contrast '[U]ninvited forms of public engagement are usually about challenging just these unacknowledged normativities' (Wynne 2007: 107). Against a backdrop of controversy over the introduction of novel environmental risks, or failure to adequately address such risks. Wynne understands 'uninvited' public participation as arising 'in response to expert-led, expert-justified interventions and misrepresentations, exacerbated by further expert-led impositions of provocative and alienating definitions of what the issues and concerns are; thus also, by misrepresentation and lack of recognition of those publics themselves' (Wynne 2007: 107).

The uninvited publics demanding to be noticed are primarily organised lay people. Deploying Wynne's concepts Peter Wehling found that organised groups in civil society proved 'to be considerably more effective than deliberations among unorganized laypeople' (Wehling 2012: 50). Wehling is careful to explain that it is important to 'distinguish collectively articulated and reflected interests from mere short-term and surface preferences such as consumer choices on markets' (Wehling 2012: 53).

Adding to the understanding of uninvited public participation Wehling reflects on the possibility of making space for the organised publics to make participation more meaningful, effective and sustainable. Two of his suggestions resonate with the practices we look at in chapter four—to make participation more long-term and to include capacity building to enable citizens to participate. With regard to environmental participation these suggestions address routine expert-informed decision making, not scientific project or one-off decision making. It is the routine expert-based environmental management decisions that affect local communities the most and they are often at the root of controversy.

In chapter four we deploy the notions of invited and uninvited publics to consider environmental participation in the context of expert-led decision making. Starting with uninvited public participation we look at the environmental justice movement that is particularly strong in the United States then we outline environmental movement expertise as an alternative that transcends the binary conceptualisation. Finally, we articulate a modified notion of invited participation through the example of community modelling that facilitates local participation in environmental governance by making scientific modelling tools, normally used only by technical experts, useful to local environmental groups.

Who Participates?

One of the topics in STS and human geography discussions of participation is the identity and characteristics of the public, or rather, publics. The plural signals that STS and many other social scientists reject the idea of a 'general public' and insist on the diversity and differentiation of publics. Human geographer Sally Eden (2017) explains that contrary to ideas about the general public there is no undifferentiated aggregate of individuals that policy makers could gauge for opinions, or who could express clear preferences for any particular governance option. Similarly, not being an environmental scientist does not unite individuals into a coherent lay public that lacks knowledge. As undifferentiated aggregated general public and as lay people lacking knowledge are two ways in which policy makers and scientists have traditionally imagined publics. These imaginaries do not withstand confrontation with practice and for a more accurate understanding of environmental publics we must take Eden seriously and acknowledge 'environmental publics are differentiated by how they relate to the environment through their environmental practices, rather than their own characteristics' (Eden 2017: 1). This directs the attention to what people do, not who they are. Eden also insists that formal, scientific, education is only one type of environmental knowledge. Talking about 'knowing publics', constituted in a wide range of practical engagements with environments, she explains that they 'are clearly very diverse: the outdoor passions of rockclimbers, anglers, birdwatchers and gardeners will vary both within and between each group, as well as differ from the indoor environmental pursuits of those watching nature documentaries on television' (Eden 2017: 36). This understanding of knowing active publics provides a starting point for thinking about the publics participating in environmental science, decision making and expertise.

Moving from Eden's discussions about environmental publics to the more specific publics participating in environmental science and decision making Mike Michael's (2009) discussion of 'publics in general' and 'publics in particular' is helpful. Reviewing the state of the art in discussions of public understanding of science Michael finds that the '"public in general" are regarded as an undifferentiated whole that is distinguished from "science in general"' (Michael 2009: 620). This is a public that can be geographically defined, such as citizens of a country, or demographically differentiated according to age or income. In contrast 'publics in particular' do, according to Michael, 'have an identifiable stake in particular scientific or technological issues or controversies' (Michael 2009: 623). Publics in particular are defined on their own terms, by their relationships with the environmental issue at hand independently of any scientific project or decision making process. Publics in particular can be of many different types, from local environmental groups, to private businesses, to tenants' organisations, home owners, local authorities, recreational groups and so on, who may all be interested in the scientific investigation of a pressing local issue, such as waste-water treatment at a particular site that could pose a risk.

The commonly used notion of stakeholder indicates an organised public with a specific interest in an environmental process or problem. These interests originate in the core activity of the organisation and it is independent of the activity in which they are invited to participate. For example, with regard to water quality in a particular river stakeholders can be anglers, water utility companies, property owners, environmental groups, local authorities and so on. Many stakeholder organisations generate professional knowledge, employing individuals who can provide them with scientific and technical information about the issue at hand. Differently from the public in general stakeholders are expected to have their own interests and agendas that prompt them to participate and to care about the outcomes of a scientific project or a decision making process.

With regard to the recruitment of participants the difference between looking for an imagined general public or publics in particular, e.g. stakeholders is significant. Recruiting members of the public in general usually starts with sampling based on demographic and other social scientific characteristics of a population defined by their geographical proximity to the environmental problem. To achieve representativity of this population is very difficult since a majority will not be interested or have the time to participate in environmental research or decision making activities. The people agreeing to participate will most likely be better educated, older and belonging to the ethnically dominant category, which poses a tremendous challenge to environmental social scientists who nourish an ambition to empower the most marginalised to improve the democratic process.

Recruiting stakeholders involves contacting organisations active in the localities affected. Scientists looking to involve stakeholders in research projects can easily identify the organised interests and invite representatives to participate in the research. It is easy to spot, for example, public agencies with regulatory responsibilities, businesses whose activities are affected by the environmental process or local environmental organisations that work to engage local communities with rivers or wildlife. Because stakeholders have their own interests in the outcomes of the participatory activity they are easy to retain throughout a research project and can be relied on to attend several events.

This discussion of stakeholders could make it seem as if they are the same as Michael's publics in particular, but this is not the case. It is important to retain the analytical specificity of the notion of publics in particular, hence, it will not be used only with reference to the organised publics participating in environmental science and decision making, but also to publics who are brought into being by an environmental matter of concern. Such publics can emerge at any time in a decision making process. They pivot on local people getting together around an issue they perceive as not being addressed to their satisfaction in expert-led decision making.

Despite critique by social scientists the idea that there is a general public that holds certain views on environmental science and governance issues is hard to shake. The preceding discussion has made clear that this imaginary public has been thoroughly deconstructed by research in STS and other fields. However, the imagined general 'public' still surfaces in conversations with scientists, experts and decision makers. Below are three, roughly drawn, commonly entertained imaginaries of the publics participating in science, decision making and expert advice.

From the perspective of environmental science the imagined public is comprised of individuals seeking more knowledge, which motivates them to participate in research activities. Research projects involving citizen science are based on this view of the public. Scientific projects that are more demanding in terms of participants' effort and role tend to add further qualifications pertaining to individual's previous experience and the time they need to be able to allocate to the project. Still, anybody who fulfil the criteria is welcome. Decision making agencies imagine a public with firmly held, but unarticulated values, that can be discovered and used to inform the decision making process. Decision making is more circumspect with regard to who is a suitable participant because representation is important.

In a democratic system everybody has a right to be heard and in environmental decision making this can translate into a desire to involve people who can be thought to represent the relevant population, for example in a demographic, socio-economic or geographical sense. People who are associated with organisations with a particular view on the issue are usually not seen as representatives for the public, but as special interest groups. Expert bodies tasked with turning abstract science into practical advice imagine their participating publics as groups defined by their relationship to the problem. Addressing actual management issues these expert bodies often work with stakeholder representatives who can provide different views on the problem and potential solutions. The imaginary general public is often seen as a problem, the procedure requires their inclusion, but they do not see things in the same way as the experts. That empirical research repeatedly refutes these imaginaries has not made them disappear and it is important to be aware of their existence to be able to challenge their unqualified invocation.

The understanding of participating publics developed in STS case studies has brought attention to the role of the practical organisation for the characteristics and identities of participating publics. Javier Lezaun and Linda Soneryd (2007) write about 'technologies of elicitation' with reference to the organisation of events to which publics and stakeholders are invited. In their view 'instruments such as the discussion group, the counselling meeting, or the citizen jury, designed to generate lay views on the issues at hand, and feed those opinions into the policy process' (Lezaun and Soneryd 2007: 279) are constitutive to the views expressed by the public. This challenges the idea of a pre-existing public opinion that can be discovered by applying the right tools which is common in environmental decision making.

The social science discussion of public participation has focussed on environmental governance (often on the national level) rather than science, but the analyses are useful for thinking about publics participating in research as well. The idea of technologies of elicitation encourages questions that cut across the distinction between science and decision making and it draws attention to the variety within each field.

What Do Participants Do?

The idea that participating publics are constituted in the technologies of elicitation used to organise participatory activities directs our interest towards practices. It becomes interesting to examine what the invited participants are expected to do and how this connects with the form and content of science and decision making. For example, are participants in scientific projects invited to carry out research tasks together with scientists, or to discuss research findings, or to listen to presentations of research results? Are participants invited to discuss possible interventions before the list of options is fixed, or when decision makers have settled on the 'best' management strategy?

Specific technologies of elicitation organise participation in time and space. We can ask what difference it makes if participatory activities are set apart from the science or decision making processes, or if participants are invited to the sites of research and decision making? Does this matter in the same way to all types of publics? Another important issue is whether there is one participatory event or a series? Intuitively we may think that longer term participation would have more impact, but what if it is arranged separately from the science or decision making and focussing on education, or if it is propelled by an instrumental rationale? The concepts introduced in this chapter do not refer to neatly separated processes, but to activities and processes that interact in complex ways making outcomes impossible to predict or control. This complexity will become clear in the following three chapters of the book as we examine different environmental participation practices in detail.

Environmental Participation in Three Dimensions

In the next three chapters of this book, I draw on my experience of doing STS research in transdisciplinary environmental projects over ten years. This is combined with information about environmental participation from case studies in STS and human geography. The distinction between environmental participation in science and decision making is reflected in chapter two and three. In chapter two the focus is on environmental science. We look at ways in which publics are invited to participate in the creation of scientific knowledge about environmental processes. In this chapter the notions of co-production, dialogue and education are used to distinguish between three major formats of public participation in environmental science.

In chapter three we turn to public participation in environmental decision making. In this chapter my experience is of a traditional social science kind, studying participatory processes in which I was not directly involved. In this chapter the notions of instrumental, substantive and normative rationales for public participation in decision making organises an examination of the practices of the West Cumbria MRWS Partnership, in existence between 2009 and 2012, and the Catchment Based Approach in water management introduced in the 2010s.

The practice-based distinction between participation in science and in decision making lets us discern activities that are designed with the ambition to connect the two. Expertise is the label applied to activities that connects science and decision making. Experts extend scientific knowledge by applying it to local environmental issues. In the fourth chapter the notions of invited and uninvited public participation provide the starting point. First, we outline environmental justice activism as an example of how uninvited publics can impact on expertise and decision making by direct action. Next, we touch on environmental organisations that have been active as long as the institutions of environmental science and decision making and developed independent expertise. Finally, we look at how scientific tools used by experts can be adapted for use by local environmental groups to facilitate public participation in environmental decision making.

The final fifth chapter in this book sums up the insights gained about the varied multitude of environmental participation practices and reflects on which issues it is important to consider in order to promote environmental participation.

References

Buytaert, Wouter, et al. 2014. Citizen science in hydrology and water resources: Opportunities for knowledge generation, ecosystem service management, and sustainable development. *Frontiers in Earth Science* 2: 1–21.

Callon, Michel. 1999. The role of lay people in the production and dissemination of scientific knowledge. *Science Technology and Society* 4 (1): 81–94.

Chilvers, Jason, and Matthew Kearnes. 2016. Science, democracy and emergent publics. In *Remaking participation: Science, environment and emergent publics*, ed. J. Chilvers, and M. Kearnes, 1–27. London and New York: Routledge.

Eden, Sally. 1996. Public participation in environmental policy: Considering scientific, counter-scientific and non-scientific contributions. *Public Understanding of Science* 5: 183–204.

Eden, Sally. 2017. *Environmental publics.* London and New York: Routledge.

Lezaun, Javier, and Linda Soneryd. 2007. Consulting citizens: Technologies of elicitation and the mobility of publics. *Public Understanding of Science* 16: 279–297.

Michael, Mike. 2009. Publics performing publics: Of PiGs, PiPs and politics. *Public Understanding of Science* 18 (5): 617–631.

Quet, Mathieu. 2014. Science to the people! (and experimental politics): Searching for the roots of participatory discourse in science and technology in the 1970s in France. *Public Understanding of Science* 23 (6): 628–645.

Raymond, Christopher M., et al. 2010. Integrating local and scientific knowledge for environmental management. *Journal of Environmental Management* 91: 1766–1777.

Reed, Mark S. 2008. Stakeholder participation for environmental management: A literature review. *Biological Conservation* 141: 2417–2431.

Stirling, Andy. 2007. Opening up or closing down? Analysis, participation and power in the social appraisal of technology. In *Science and Citizens: Globalization and the challenge of engagement*, ed. M. Leach, I. Scoones, and B. Wynne, 218–231. London and New York: Zed Books.

Wehling, Peter. 2012. From invited to uninvited participation (and back?): Rethinking civil society engagement in technology assessment and development. *Poiesis & Praxis* 9: 43–60.

Wynne, Brian. 2006. Public engagement as a means of restoring public trust in science—Hitting the notes, but missing the music? *Community Genetics* 9: 211–220.

Wynne, Brian. 2007. Public participation in science and technology: Performing and obscuring a political-conceptual category mistake. *East Asian Science, Technology and Society: An International Journal* 1: 99–110.

CHAPTER 2

Public Participation in Environmental Science

Abstract This chapter focuses on the participation of publics in environmental science. The STS classification—co-production, dialogue and education—of different ways in which science and lay people relate, is introduced as an analytical framework. This framework organises the discussion of environmental participation practices in three clusters: co-production projects, stakeholder deliberation and citizen science. Examples of each cluster are presented, demonstrating the diversity of environmental participation in scientific research. Finally, I reflect on how the diverse practices could challenge the framing typology and raise some issues that seem to be left out of most discussions of environmental participation.

Keywords Environmental science · Co-production · Stakeholder deliberation · Citizen science

ENVIRONMENTAL SCIENCE AND ITS PUBLICS

As discussed in Chapter 1 public participation in environmental science has several origin stories. One story looked to the radical science movements of the 1970s which gathered natural scientists who were critical of the uses of science in war and destructive exploitation of nature (Quet 2014). Another story starts in the late 1990s when scientific organisations

© The Author(s) 2020   23
C. Landström, *Environmental Participation*,
https://doi.org/10.1007/978-3-030-33043-9_2

identified a decrease of trust in science (Wynne 2006). Further origin stories could highlight the political ambitions of research funding bodies to
involve concerned publics in the creation of knowledge that is used to
manage environmental risk. Regardless of which origin story we tell environmental scientists today are getting used to view engagement with the
public as a regular feature of research projects.

There are numerous ways in which scientists can relate to publics, in
this chapter we use Michel Callon's (1999) threefold typology to organise
the discussion of participatory practices. Callon's distinctions between 'co-
production', 'dialogue' and 'education' can be understood to pivot on the
impact public participation is allowed to have on the scientific research. The
co-production model is the most radical, involving scientists and lay participants as equals, working together to create new knowledge. In Callon's
second model, 'dialogue', the interaction takes place outside of the research
process. Scientists and lay participants get together to meet and tell each
other about their respective perspectives on the issue. Aiming for exchange
of knowledge, ideas and opinions 'dialogue' does not intend to change
the science, but to align it with the objectives and needs of its publics.
The objective of dialogue is primarily to improve the usefulness of new
scientific knowledge by providing scientists with a better understanding of
what their publics want and need. The third model for science's interaction
with publics that Callon named 'education' needs further elaboration to
resonate with environmental participation. Assuming that Callon's intent
was not to talk about formal science education in schools and universities,
we can think of education as referring to the practices in which scientists
communicate knowledge with a specific content to a particular public. In
the context of environmental participation this can apply to the instructions provided in citizen science to enable participants to undertake the
tasks involved.

Before proceeding to practical examples it is important to note that in
the context of scientific research public participation is a research method.
The objective of scientists who invite publics to participate is always to
further the objectives of science.

Co-production: Environmental Competency Groups Doing Science Differently

This discussion of environmental participation in research starts with co-
production because it is how I came into the field. Three local 'competency
groups' conducted in two transdisciplinary projects, is a limited experience,

but it suffices to outline some key aspects of co-production in practice. When I took up a post-doctoral position in a project led by the human geographer Sarah J. Whatmore on flood risk management, participatory research was new to me.[1] As an STS researcher with a background in the subject Theory of Science and Research my exposure to lay publics had thus far been minimal. Thankfully the project began with educating the researchers.

Attending the initial team preparation weeks with the other project scientists I learned that the flood risk project was the first trial of the competency groups method in the UK. I understood the overarching aim to be that of making the knowledge of local residents, affected by an environmental problem, contribute to the creation of scientific knowledge about the issue. The ambition was to generate new perspectives on local problems to extend scientific understanding in ways that could make it possible to assess a broader range of flood management options. In the team training sessions, we learned from each other, the natural scientists explained what their research questions were and about how they used computer models to analyse flooding.

Engaging local publics in flood management was not new, but competency groups were unique. A key feature was to create a space for local participants to interrogate and critique expert knowledge, with the support of scientists. This is an expression of the critical STS underpinnings of the methodology. Competency groups emphasise the right of citizens to disagree with institutional policies based on expert knowledge in a way that echoes the understanding that scientific knowledge depends on rigorous critique within scientific communities. All knowledge claims presented in a competency group are open to query. Scientists' knowledge claims are not accepted just because they are thought to be 'scientific'. Equally, experience-based knowledge claims are not accepted at face value. Competency groups keep the environmental problem addressed open, allowing the group to explore it.

The work in competency groups centres on object. Working with things like maps, or photographs, is much more inclusive than working with numbers or scientific accounts. This is also reflected in the idea that groups should aim to make something together that can be delivered to the wider

[1] Understanding Environmental Knowledge Controversies: The case of flood risk management 2007–2010, funded by the Rural Economy and Land Use (RELU) programme, PI professor Sarah J. Whatmore, Co-Is professor Stuart Lane and professor Neil Ward.

community and enable them to learn through the competency group's findings.

Participants are asked to view joining a competency group as an individual commitment and understand that all input to the group is considered as personal contributions. Competency group members do not participate as representatives of institutions, organisations or family and friends. This is important since the new knowledge created could make people change their minds about an issue and this could put them in a difficult position if they were committed to speak for others or promote a set agenda.

In the team training sessions we developed a format for the competency groups that we would arrange. The core activity would be six scheduled meetings over the course of one year. The three teams at different universities would get together before each competency group meeting to prepare the activities of the group. These preparations would build on the previous meeting of the competency group and the input from all group members (university researchers and local people) between the scheduled meetings. While it is not possible to know, or desirable to control, the content or outcomes of competency groups it is absolutely necessary to prepare practical tasks.

Obviously, the first meeting could not rely on input from previous meetings or local participants, and there are some elements that are needed in order to get going. Firstly, all members have to be introduced. One way to make presentations collaborative is to place people who do not know each other in pairs, ask the two to present themselves to each other and then let everybody present the person they talked to. This is a good way to make scientists talk about themselves rather than their scientific knowledge if they are paired up with local participants. Further activity in the first meeting is to find out how everybody relates to the local environmental problem. Another task is to agree on the way the group will function, to collaborate on an ethical protocol and make clear what all group members can expect from each other. Finally, the first meeting generates suggestions for what to focus on in the second meeting. This was the key element of the plan with which we embarked on the first competency group in North Yorkshire.

Environmental Participation Changing Flood Risk Management

The very first trial with competency groups was undertaken in Pickering, a market town in North Yorkshire.[2] The project leads had identified this as a locality of interest to both social and natural scientist. At the time there was long-running acrimonious controversy about how to address flood risk. The responsible authority, the Environment Agency (EA) had given up after the strategy they had proposed lost funding and the knowledge underpinning it was challenged by local people. As a consequence Pickering that had suffered several flood events, was left without any flood risk reduction measures.

The project team of seven arrived in Pickering in late spring 2007. One of the social science sub-teams recruited people who were interested in opening up the knowledge about flooding in the town. An important point is that we never promised that the research would provide a solution to the problem. We could not know if we would be able to find a better way to approach the problem, science is uncertain, and it is critical to communicate this to people who consider participating in a research project. In the time that passed between advertising and interviewing potential participants Pickering was hit badly by unexpected summer flooding. This event most certainly contributed to raise interest in participating and to a sense of urgency of finding some way to mitigate flood risk. We successfully recruited local residents who were interested in exploring the problem.

The first session worked exactly as we had hoped, the research team from the universities of Oxford, Durham and Newcastle and the local participants got talking to each other as equals, with differently constituted knowledge. Discussing flood events in the group helped the scientists understand how floods behaved in Pickering and the local participants learned about how scientists investigate flooding with the help of computer models. An agreement to focus on computer modelling emerged to the surprise of the natural scientists. They had expected local people not to be interested in modelling as it is thought to require expert technical skills. However, the local participants were fully aware of the role modelling played in flood risk management and saw an opportunity to learn about and use the tools that the powers that be had access to.

---

[2] This project provided subject matter for several academic articles, for example, Landström et al. (2011), Lane et al. (2011), and Whatmore and Landström (2011).

Computer modelling had been at the centre of the failed attempt to manage flood risk by building flood walls along the river in the town centre. This was opposed by a group of local people who challenged the validity of the technical report. The belief in flood walls as providing the only effective mitigation option, firmly held by most parties in the local controversy, was based on modelling results presented in a consultant report commissioned by the EA. Having seen modelling playing such a key role in flood risk management and experienced the difficulty of challenging it with knowledge created in other ways, the local participants wanted to address modelling critically.

The importance of computer modelling for the situation in Pickering informed the second meeting in which we used maps with modelled flood outlines. We asked the local participants to mark, with coloured pens, where they had seen flooding or been told that it had occurred. In this way we could query both the technical expert claims and local experience. As local participants compared their records and memories new collective experience-based knowledge about flooding in Pickering emerged, together with a map providing a representation of actual flooding in contrast to model outputs.

Querying model projections in this way was important both for the articulation of collective knowledge about flood patterns in Pickering and for making clear to the scientists what information they did not have. The modelling was based on instrument generated measurements and these will always be limited. In contrast, the detail provided by local participants was critical to understanding flooding in Pickering in the depth required to model it differently and potentially find other options to reduce flood risk. The natural scientists in the competency group had an interest in exploring the possibility of changing land use to mitigate flood risk. The gap between modelled and observed flooding in Pickering encouraged them to try different modelling approaches. After a while, they arrived at the conclusion that they had to invent something new to address the questions arising in the competency group.

For the third meeting the natural science modellers had prepared a first, rough version of a new computer model that coupled hydrology with hydraulics and used a routing algorithm to simulate the movement of water across the landscape in Pickering Beck. This 'bund model' allowed the group to explore the potential flood mitigation effect of placing several, small dams upstream in the river to reduce the amount of water arriving in the town centre at any one time. This model became the focus of group activity in the fourth and fifth meetings and in the final meeting we planned an event in the Town Hall where the new model and the associated proposal for natural flood risk management was presented to the wider local public.

The competency group in Pickering was a spectacular success in both scientific and local impact terms. The local participants in the group continued to promote the upstream bunding proposal and managed to turn it into a national demonstration project called 'Slowing the Flow' and eventually measures originating in the group's work were implemented in Pickering Beck.[3] This outcome was serendipitous, there is reason to believe that the competency group acted as a catalyst in a situation with a unique set of conditions. We thought that it would be unusual for competency groups to have such concrete impact, which was what we found in the subsequent two competency groups.

Complementary Co-production in Uckfield and Marlborough

The project organising the competency group in Pickering had been designed to run two such participatory activities. The second location was Uckfield in East Sussex, where the situation was very different. Although badly flooded in 2000 there had been no flooding for nine years when we arrived and the controversy about risk reduction had abated. We managed to recruit participants for the first session, and we found that a driver for several local participants was to learn about flood risk in order to develop strategies for the future. Some of the local group members were also active in local environmental groups and could contribute the knowledge of other collectives. Other group members had professional experience of working with flood management in the locality and could contribute unique insights.

In the second session we worked with maps to relate experienced floods with modelled floods. Some of the initial participants left the group before this session, but other local people had been invited by the remaining group members. The local participants had identified people with relevant knowledge and asked if they could invite them. Changes of participants went on through all meetings and was a feature that prompted the group to explore a range of different issues. To the natural scientists this meant that a range of modelling possibilities could be explored in depth and eventually the

---

[3] The UK Forestry Commission provides an account of the demonstration project at https://www.forestresearch.gov.uk/research/slowing-the-flow-at-pickering/.

Uckfield group contributed significantly to the development of a computer model called 'Overflow' that was based on the bund model, but much more sophisticated.[4]

The Uckfield competency group co-produced scientific knowledge in a way that is what this participatory approach aims for. A range of differently constituted local knowledge shaped the process, defined which questions it was important to explore, corrected assumptions about empirical conditions and scrutinised the modelling outputs. It had no discernible impact on local flood risk management.

The third environmental competency group (ECG)[5] was undertaken in a different scientific project and instead of flooding the topic was drought. Gathering in Marlborough this ECG was organised as part of a multidisciplinary project investigating drought in the UK.[6] The River Kennet, which joins the River Thames in Reading, was interesting for several reasons. It is a chalk stream and very sensitive to drought, it ran dry in the summer of 2012. There has also been a controversy raging for decades over the amount of water that the utility company Thames Water abstracted from the Kennet catchment. At the time of the ECG a solution had been agreed and abstraction from the Kennet ground waters would decrease significantly.

The local advertising attracted enough local participants to start the Kennet ECG. There was some concern in the academic team that drought would appear remote from everyday concerns, but we found that there were several local groups that kept water management on the local agenda. Although many interested people were members of local environmental organisations they agreed to participate in the ECG as individuals wanting to explore the issues, not as representatives for any organisation.

This group progressed as planned, the participants returned for all sessions and because we had three different types of scientific modelling expertise available in the group we could examine different processes. A key focus was the water quality in the Kennet, we explored how climate change

[4] This model is presented in Odoni and Lane (2010).

[5] Producing a multitude of accounts of the work in Pickering and the method trialled we realised that it was important to signal that we did not know whether the approach would be useful for participation in research on other than environmental issues, hence we added the qualifier environmental to competency groups.

[6] Professor Jim Hall was PI for the MaRIUS (Managing the risks, impacts and uncertainties of drought and water scarcity) and it involved Co-Is at University of Oxford, University of Bristol and Cranfield University.

increasing extremely low and high flows could affect it. We also looked at how future changes of the amount of water abstracted for human use could impact on the flows and thus, on water quality. And, we explored the potential of changing agricultural practices to reduce the amount of runoff causing water quality problems. Through this collective modelling the group co-produced new knowledge about the ways in which local farming impacts on water quality in different conditions. The scientists also gained deeper understanding of the limitations of their modelling, learning that, for example, the scale of water resource models makes rivers like the Kennet almost invisible. The social scientists learned a lot about how UK water management looks from the local level, there was a sense of fragmentation due to separation of policies for flooding, drought and water quality. The River Kennet ECG co-produced a report with the intent of bringing water issues and river status to the attention of the local council's work with developing an Area Neighbourhood Plan.[7] This was a document that would guide local development, with regard to the siting of housing, industry and other built structures, the Kennet ECG insisted on the importance of considering the impacts of local planning on the river.

Lessons on Co-production

The three examples of co-production clarify some things about this format of environmental participation. Firstly, the lack of control. In all cases the ideas we in the project team had about what the groups would do changed significantly in the course of the collaborations. This is not surprising, in any true collaboration the sum will be different from the parts. If the participating lay people are treated as equals the co-production will take on its own dynamic. In order to be genuinely open-ended it is only possible to decide the format of the activity beforehand, differently from any scientific project it does not have a given discursive context that defines roughly what questions can be investigated and in which field answers may be found. The unpredictability brought about by involving people with very different non-scientific experiences and interests can make research turn in any direction. If a project design does not include the possibility for a participatory project to develop its own agenda, scientists may still gain new knowledge and the lay participants will learn more about the issue at

---

[7] Kennet ECG (2017).

hand, but there will not be co-production of knowledge. The lack of control on behalf of scientists also make co-production high-risk in terms of producing useable results, it is impossible to predict what, if anything, will become significant. In the Kennet ECG much attention was directed to water governance issues, we made use of scientific models to consider the possibility of acting in new ways with regarding who would be involved in which decision making processes. There were some very interesting lessons about decision making agency for the social scientists participating in the ECG, but no challenges to existing natural science knowledge claims.

Co-production can be highly rewarding for all involved parties and have radical consequences for the way scientists understand and represent environmental processes. However, it is also uncertain, slow and costly, it can deliver big gains in knowledge, but risks not making much difference to science or society. Hence, it makes a lot more sense to most research projects to use the more goal-oriented form of public participation that I call 'stakeholder deliberation'.

Scientists Deliberating with Stakeholders

The participatory approach that I call stakeholder deliberation is a much more common way for environmental scientists to involve non-scientists. The objective is usually to align research with the explicit knowledge needs of stakeholder organisations. This way of working brings scientists and stakeholders together at specially designed events while the research is going on but separate from the research activities.

As discussed in Chapter 1 stakeholders are organised publics in particular, defined through a pre-existing relationship with the environment that provides them with an explicit interest in the research. Stakeholder organisations are represented by individuals who participate in the activities arranged by the scientists. These representatives often have a personal interest in the issues, but that is not the reason for them being invited to participate in research projects, they are there to speak for institutions, corporations or other organisations. This is a narrow use of the term stakeholder, often scientists refer to all people involved with an ongoing research project who are not researchers as stakeholders.

In practice stakeholder deliberation centre on presenting, discussing and aligning scientific and societal actors' interests. The scientists and the stakeholder representatives meet to discuss the research, to articulate their interests and to agree on how to contribute to each other's missions. It has become very common for environmental research projects to include stake-holder deliberation in the hope of increasing the societal uptake of research findings. This can be interpreted as a reaction to a perceived failure of environmental science to establish and maintain clear connections with society.

Environmental science has always been expected to contribute knowledge that is of use to the management of environmental problems and risks. However, growing in extent and evolving theoretically over time, much environmental science today is what would be considered basic, or curiosity-driven, rather than prompted by societal needs. The curiosity of scientists and the theoretical discourses of academic science are often quite remote from the practical knowledge needed by actors in society. Demands for stakeholder involvement can be viewed as a reaction to this 'scientisa-tion' of environmental research.

Two of the projects I have worked in have had regular stakeholder deliberation events. This form of environmental participation finds favour with scientists for many reasons. One is the minimum impact on the conduct of research and another the possibility of generating documented interaction with institutions and organisations outside of science and research, an explicit relationship with decision makers. Such documentation provides evidence that the research has been communicated to relevant parts of society.

Deliberating with Stakeholders in a National Research Programme

The four projects in the NERC (Natural Environment Research Council) programme investigating drought and water scarcity in the UK brought together most of the country's stakeholders interested in processes affected by drought. All four projects involved stakeholder deliberation and closer examination shows what the notion means in practice.

The project focussing on drought impacts, invited stakeholders to join an advisory group. This stakeholder advisory group met regularly with the project leads to discuss the knowledge produced in the project and the knowledge needs of the stakeholder organisations. The stakeholder representatives were also invited to all public events organised by the project and project researchers were encouraged to interact with the stakeholder representatives in relation to their specific tasks.

The stakeholder representatives came from different organisations, including a water utility company, a regulatory agency, a public body for environmental protection and an engineering consultancy. Interviewing these individuals, I found that they all had natural science and/or engineering backgrounds and a personal interest in water issues. However, they understood their roles in the research project to be to represent their organisations' interests in keeping up to date with the latest scientific research and findings.

The scientific/technical background of the stakeholder representatives adds intrigue to the relationship between stakeholders and research scientists. These individuals were not lay people in relation to drought research, but they were not university researchers. They represented what the project scientists considered legitimate societal interests. The project scientists viewed the stakeholder organisations as representing the 'society' that would benefit from the knowledge generated in environmental science. The main route for the new scientific knowledge into drought management was to be the stakeholder representatives.

The relationship between stakeholder organisations and scientific project was understood as mutual, in addition to telling the scientists what they needed more knowledge about, which types of scientific tools they found useful and what they would like the research to deliver the stakeholder representatives would contribute information about drought that their organisations had. The scientific project would make use of the stakeholder generated data as well as orient their research questions towards stakeholder issues, as well as towards scientific discourses and make sure that research outputs were delivered in a format that would be of use to the stakeholders. For this multidisciplinary project, the stakeholder needs were also a way to align research teams with different disciplinary specialisms. The different university teams were geographically dispersed in a way that mostly corresponded with disciplines and work tasks. Addressing stakeholder questions that cut across existing divisions of labour provided rationales for new collaborations among the researchers in the project.

The stakeholder meetings were organised as half-day seminars, taking place in the offices of the lead scientific institution. A selection of project researchers, based on which team had the most relevant findings to present attended, together with project leads. The researchers, mostly post-doctoral researchers working full-time in the project, had prepared presentations tweaked to encourage stakeholder input. Each presentation was discussed in detail, ample time was allotted to comments from the stakeholder representatives and project researchers from other teams, to consider the research from their different perspectives.

The exchange of scientific information among natural scientists and professionals that took place in the stakeholder meetings of the drought project seems very distant from the idea of public participation. However, this way of linking environmental science with actors in society is very common, it fulfils the ambition of scientists to engage with key users of knowledge and grants institutions direct access to science. If society is understood as represented by institutions and formal organisations this type of participation satisfies the demand on science to serve society. From an STS perspective this is problematic, as it excludes actors who are not formally organised and alienates stakeholder representatives with different disciplinary backgrounds. The stakeholder advisory group in the drought project started out with a diverse range of individuals representing different types of organisations, but over time the representatives with educations less similar to the project scientists and organisations less focussed on water resources dropped out. In the end only the scientists and the stakeholders who had been collaborating before the project started remained. This is not in principle a problem for participation in environmental science that aims to improve scientific knowledge, not democratic representation.[8] However, it is important to recognise the very limited scope of this type of environmental participation.

Stakeholder Deliberation in an International Project

Stakeholder deliberation was also a feature of a very different project that brought together social scientists from eleven countries to study sociotechnical challenges in radioactive waste management. Funded by Euratom this project had a close relationship with the national agencies responsible for implementing radioactive waste strategies in the respective country.[9] The stakeholder representatives coming from these organisation were mostly technical experts with natural science and engineering backgrounds who wanted to understand the social aspects of the issue in order to promote the implementation of geological disposal (GD) in their countries.

In this project, the stakeholder representatives joined the project meetings at a dedicated time and were treated to presentations of the work of

---

[8] The lack of diversity among the actors does, however, contribute to a lack of innovation in UK drought management according to Grecksch (2018).

[9] For an account of this project see the final report Kallenbach-Herbert et al. (2014).

the social science researchers. The differences in educational backgrounds of the project team and the stakeholders was mitigated by the fact that the stakeholders had practical experience of dealing with their own national stakeholder groups in the national decision making processes. Since there are very few countries that have so far managed to do anything more than research and deliberation with publics with regard to GD the technical experts had had several years of experience with the sociotechnical challenges and were well equipped to understand the perspectives and analyses of the social scientists, although they did not necessarily agree with our conclusions.

Discussions at the stakeholder seminars were animated, particularly since the social scientists were very open with not being interested in promoting the implementation of GD, but in understanding the complex sociotechnical dynamics involved in the siting of geological disposal facilities (GDFs) in specific localities in the different countries with their unique histories and political cultures.[10] In this case, the stakeholder participation involved representatives of organisations with a very specific interest in the social science project—they wanted to know how to implement GD. Stakeholder deliberation was critical for the research in this project since it provided the research team with knowledge about how the agencies directly responsible for the management of the environmental risk understood the issue.

The experience of the transnational project on radioactive waste challenged me to think beyond local organisations in relation to stakeholder deliberation in environmental science. I would not have thought that stakeholder participation would make sense at the transnational level since environmental problems are primarily addressed in national policies, but geological disposal of radioactive waste is a unique issue. There is only one responsible agency in each country and they usually plan for one national GDF and the implementing agency in each country tend to deal with one local community if a potential site has been identified. For social scientists to be able to do any type of comparison cross-national collaboration is necessary and given the nature of the issue it would be very difficult for a social science project to get access to the implementing agencies without the reciprocity and transparency that stakeholder deliberation provides.

Comparing my experience of stakeholder deliberation in practice with what has been discussed in the literature it is necessary to remember that

---

[10] For an example of analyses generated in this project see Konopasek et al. (2018) and Landström and Bergmans (2014).

the notion of stakeholder used here is much narrower than what is usual. However, this was intentional since it is important to note that in many environmental science projects the stakeholders are experts working in private and public organisations affected by the issue subjected to research. By clarifying why some forms of stakeholder deliberation works we can understand why others are more precarious. For example, the often discussed, problem of 'participant fatigue' is less relevant when the stakeholder representatives are professionals who attend research-related events as part of their work. In contrast, participant fatigue is a clear risk when stakeholders are not familiar with scientific and technical perspectives on an issue before participating. The language and the timelines of scientific projects make them difficult to engage with. At deliberative events people without scientific background are often being talked at by scientists, and they do not understand why they were invited or what they are supposed to do. Lay participants who are treated as professional stakeholder representatives often feel that their contributions are not taken seriously, and they don't understand and are unable to use numerical research outputs. Negative experiences of participating in environmental science projects will make people cautious of getting involved again. In contrast citizen science can be much more satisfying for participants who may stay involved for a long time.

Citizen Science—Mass Participation

Citizen science is probably the best established and most wide-ranging format for environmental participation in science. The term has been used in different ways, but here it is deployed with reference to public participation in environmental research that comes about because scientists need people to collect or process information to an extent that they could not do themselves. Citizen science is a type of participation to which everybody who is interested can be invited. Recruitment can be done by advertisements placed where the scientists initiating a project believe that people with an interest might see them, ranging from general news media to specialist online chat forums. In citizen science the aim is not to access the knowledge of participants or tailor research findings to their needs, but they are invited to collect information useful to the scientists, hence, it does not matter who they are.

Participants in citizen science projects can be widely dispersed working individually with, for example, counting the number of insects sticking to a car number plate at a particular day in a specific locality which will provide scientists with numerous point observations with geographical reference. Participants can also be recruited as a group that will carry out a specific task, for example, a school class can spend a day searching for invasive species of crab in a coastal location.

While the main objective of citizen science is to recruit large numbers of people to carry out simple tasks that cannot be automated, and do not require specialist knowledge, it also has educational aspects. Most concretely the participants must receive some training and instruction to enable them to carry out the task correctly, which inevitably increases their knowledge about the environment. In many cases participating in citizen science also increases the participants' already existing interest in the topic and encourages them to learn more, sometimes this leads to further engagement with the scientific project. Citizen science promotes education as a side effect of training interested individuals to carry out new tasks and, in many cases, this is furthered by feedback arrangements that allow participants to see what their information is used for. In the environmental sciences there is a multitude of citizen science projects involving people in varying types of activities, some involve minimal action on behalf of participants and no face-to-face interaction.

Spatially Distributed Mediated Citizen Science

In the project on drought and water scarcity discussed above the climate science modelling was done with the citizen science computing network Weather@Home (W@H). The climate modellers in the project submitted requests for model runs with the parameters they had defined, to the technical experts responsible for the distributed computing in the citizen science project. When the runs had been completed the climate scientists received the output information in the form of numerical data that they made sense of by applying algorithms capturing the weather patterns of interest to the project.

W@H comprises regional climate modelling experiments undertaken within the climateprediction.net global citizen science initiative. Climateprediction.net traces its origin to an article by the UK climate scientist Myles Allen about 'do-it-yourself' climate modelling (Allen 1999). Allen's idea was twofold, on the one hand, he found the public understanding of climate change and climate modelling lacking and wanted to get more people interested and informed. On the other hand, climate change ensemble modelling required more computer power than what the scientists could muster on their own at the time. Inspired by the Seti project (that searches for extra-terrestrial life with the help of thousands of people letting the project use a small amount of processing power on their home computers to scan radio-telescope data) Allen suggested that climate modellers could do the same. In 2002 a group of climate scientists in UK universities received funding to develop this idea. It went well and the first scientific climate modelling article based on this method was published in 2005 (Stainforth et al. 2005). Since then climatepredicition.net has become firmly established as an infrastructure for climate modelling. The regional models used in W@H were launched in 2010 and made it possible to increase the resolution of ensemble modelling to consider things like drought in the UK.

To become a participant in climatepredicion.net you download the software from the website (http://www.climateprediction.net). Participating requires minimum input on behalf of the public, however, early on the scientists learned that participants wanted more, they were genuinely interested in climate modelling. A participant message board was created, initially to address potential technical problems, but quickly extending with discussions of climate modelling as such. Today there are several message boards on specific topics and a twitter feed providing updates on the scientific work, as well as a Facebook page allowing for even more interaction.

W@H is peculiar in that it does not require much interaction between scientists and participant, actually it is the computers that participate. However, there are many other large-scale citizen science projects in which the participants can get much more involved, both with the scientists and the scientific content. The Swedish 'Artportalen' (species portal) is one example.

Artportalen is an online system for lay people to report observations of species of plants and animals (https://www.artportalen.se/). Starting in 2000 the number of reported observations reached 67 million in 2019. The system has 16,000 users and on average receives a new observation every fifth second. 'Artportalen' has become a national environmental monitoring system, keeping track of changes in the numbers and distribution of both common and uncommon animal and plant species. These data are used by scientists and decision makers to study and protect the Swedish natural environment. The website provides access to scientific analyses and publications, and there is also a contributor-generated photo gallery. Artportalen involves active public education, scientists present new knowledge in a language and formats comprehensible to the publics who contributed information.

Large-scale citizen science projects, such as W@H and Artportalen, have emerged as new scientific infrastructures that allow environmental scientists to analyse processes that would lie far beyond the reach of one project, or even a network of scientists. However, there are also many small, time-limited and local citizen science projects.

Local Citizen Science Monitoring

Working in research projects on water in the UK I have met with many people involved with local rivers trusts. These non-profit stewardship groups are often keen to take part in different practical research activities, the volunteers they organise enjoy being outdoors doing things in and by the local river. One citizen science activity that many rivers trusts get involved with is the 'river-fly kick'.

Kick sampling is a method for monitoring water quality in freshwater rivers by collecting macroinvertebrates. Rivers trusts volunteers visit a site at regular intervals to collect samples and record the data. Local rivers trusts engage with this data collection method to different degrees, one group that has made it a key feature is the Ribble Rivers Trust in Yorkshire (ribbletrust.org.uk).

Commencing in 2008 the riverfly monitoring by the Ribble Rivers Trust volunteers now covers 100 sites. To participate you sign up to attend a one-day training workshop that prospective volunteers can get a sense of through a short video from such a workshop arranged in 2016. The kick sampling activities by individual rivers trusts are linked together by the Riverfly Partnership; a network of organisations, including local environmental groups and scientists, engaged with water quality in UK rivers (http://www.riverflies.org/). Initiated by anglers and now adopted by other groups, the riverfly monitoring initiative is described as a 'bottom up' citizen science project. The threefold aim is to: 'protect the water quality of our rivers; further the understanding of riverfly populations; and conserve riverfly habitats'. The Riverfly Network is hosted by the Freshwater Biological Association, an independent scientific organisation. The FBA has promoted freshwater science since 1929 and has over time expanded their scope to appeal 'to those with a general interest in freshwater biology, including young people and amateur enthusiasts' in addition to scientists (www.fba.org.uk/what-we-do).

Riverfly monitoring by kick sampling is a paradigmatic example of environmental participation in local citizen science. It fulfils the dual objective of creating a scientific species inventory and educate the participating volunteers. In addition, it is a practical task that can help local organisations to advertise and recruit new members to a hands-on activity that is scientifically valuable. It also provides an opportunity to raise awareness about the river in the community and to teach local communities about the river ecosystem and about water quality.

Citizen science is environmental participation in science for everyone, participating does not demand more education or skill than what any interested person can acquire from a training session. Interestingly citizen science can have significant influence on science. The scale of some citizen science projects could not be replicated with other means and by undertaking citizen science environmental researchers can address questions in ways that would not otherwise be possible.

CONCLUDING REFLECTIONS ON ENVIRONMENTAL PARTICIPATION IN SCIENCE

Drawing on my own experience of working in transdisciplinary projects I have argued that the most radical form of environmental participation is co-production. Deliberation with stakeholders is more formal and often involves professionals representing organisations. Regardless of the radical impact of co-production arguably the environmental participation that makes most of a difference to environmental science is citizen science.

The threefold conceptualisation of the roles of lay people in science—co-production, dialogue and education—was useful for organising the practical examples based on what the lay participants did. In the co-production example, environmental competency groups, lay participants took part in research activities, shaped research questions and processes and produced outputs together with the scientists. At least some of the time, at other times the local participants learned about how scientists calculate flood risk and how institutions did cost-benefit analysis of different prevention measures. On close inspection activities designated co-production also involved other modes of interaction, but not necessarily in the form of scientists acting in unison in relation to a singular public. There was also education of the scientists on how water behaved in specific localities and on the challenges posed by fragmented policy. Occasionally subsets of the transdisciplinary research team deliberated with institutional stakeholders in the locality.

The dialogue model, used with reference to practices of stakeholder deliberation, also turned out to be less contained. An interesting question is to what degree exchange of data involves co-production. In the drought project stakeholders held important, proprietary measurement data about the amount of water available under different conditions. Sharing this data with the research project enabled the modellers to do things that they would not otherwise have been able to. In a reciprocal movement the data generated by the project was made into a publicly accessible resource at the end which was appreciated by the modellers in the stakeholder organisations. This exchange resembles co-production with regard to the impact on the research.

Not surprisingly citizen science is no simple tool for education. W@H generated data that was the driver for all the modelling in the drought project. The climate event sets that all modellers in the project were using to ensure that they worked with the same environmental phenomenon, was generated through regional modelling with this citizen science initiative. In more hands-on citizen science activities scientists can also find themselves in more co-production relationships with local participants who can tell them a lot about how to avoid sampling errors due to particular circumstances. Environmental participation in the form of citizen science also has significant impact on science and governance. One example is the Swedish species data base built on the observations recorded by lay people, interested citizens with very varying degrees of specialist knowledge, ranging from birdwatchers to dogwalkers with a mobile phone app. Curated by scientists this data base has become the source material for scientific research in ecology and adjacent fields, and it is used to develop environmental regulation and policy.

The thought that comes to mind is that perhaps the resource intensive, time demanding co-production projects impact more on the science–society relationship and the ways in which environmental scientists and local people relate, than on scientific knowledge as such. In contrast large-scale citizen science, that invites everyone to take part, has significant impact on environmental knowledge by the collection of data with much higher spatio-temporal resolution than what would otherwise be available to scientists.

References

Allen, Myles. 1999. Do-it-yourself climate prediction. Commentary. *Nature* 401: 642.

Callon, Michel. 1999. The role of lay people in the production and dissemination of scientific knowledge. *Science Technology & Society* 4 (1): 81–94.

Grecksch, Kevin. 2018. Scenarios for resilient drought and water scarcity management in England and Wales. *International Journal of River Basin Management* 17: 219–227.

Kallenbach-Herbert, Beate, Bettina Brohmann, Peter Simmons, Anne Bergmans, Yannick Barthe, and Meritxell Martel. 2014. *Addressing the long-term management of high-level and long-lived nuclear wastes as a socio-technical problem: Insights from InSOTEC*. University of Antwerp. ISBN: 9789057284779.

Konopasek, Zedneck, Linda Soneryd, and Karel Svacina. 2018. Lost in translation: Czech dialogues by Swedish design. *Science and Technology Studies* 31 (3): 5–23.

Landström, Catharina, and Anne Bergmans. 2014. Long-term repository governance: A socio-technical challenge. *Journal of Risk Research* 18 (3): 378–391.

Landström, Catharina, Sarah J. Whatmore, Stuart N. Lane, Nicholas A. Odoni, Neil Ward, and Susan Bradley. 2011. Co-producing flood risk knowledge: Redistributing expertise in 'participatory modelling'. *Environment and Planning A* 43 (7): 1617–1633.

Lane, Stuart N., Nicholas A. Odoni, Catharina Landström, Sarah J. Whatmore, Neil Ward, and Susan Bradley. 2011. Doing flood risk science differently: An experiment in radical scientific method. *Transactions of the Institute of British Geographers* 36 (1): 15–36.

Odoni, Nicholas, and Stuart Lane. 2010. Knowledge-theoretic models in hydrology. *Progress in Physical Geography* 34 (2): 151–171.

Quet, Mathieu. 2014. Science to the people! (and experimental politics): Searching for the roots of participatory discourse in science and technology in the 1970s in France. *Public Understanding of Science* 23 (6): 628–645.

Stainforth, D. A., T. Aina, C. Christensen, M. Collins, N. Fauli, D. J. Frame, J. A. Kettlebourough, S. Knight, A. Martin, J. M. Murphy, C. Piani, D. Sexton, L. A. Smith, R. A. Spicer, A. J. Thorpe, and M. R. Allen. 2005. Uncertainty in predictions of the climate response to rising levels of greenhouse gases. *Nature* 433: 403–406.

The River Kennet ECG. 2017. Active Water Resilience: Incorporating local knowledge in water management of the River Kennet catchment. Report of the 2015–16 River Kennet Environmental Competency Group. Published by ECI, ISBN: 978-1-874370-66-6.

Whatmore, Sarah J., and Catharina Landström. 2011. Flood-apprentices: An exercise in making things public. *Economy & Society* 40 (4): 582–610.

Wynne, Brian. 2006. Public engagement as a means of restoring public trust in science—Hitting the notes, but missing the music? *Community Genetics* 9: 211–220.

References

Public Participation in Environmental Decision Making

Abstract Examining environmental participation in decision making this chapter begins with a clarification of the term environmental decision making in relation to management and governance. Then I outline a conceptual framework based on the commonly used notions of instrumental, normative or substantive rationales for involving publics. The main part of the chapter is dedicated to two practical examples. One is the West Cumbria MRWS Partnership that spent three years examining technoscience knowledge and establishing the local community's view on the possibility of siting a geological disposal facility (GDF) for radioactive waste in the area. The second example is the Catchment Based Approach (CaBA), a policy involving local communities in river management. In the final section the examples are considered in relation to the notion of technologies of elicitation.

Keywords Decision making · Governance · Geological disposal · River management

INTRODUCTION

In this chapter we examine environmental participation in decision making. In Chapter 1, I explained that the pragmatic distinction between participation in environmental science and decision making is important for the

© The Author(s) 2020   45
C. Landström, *Environmental Participation*,
https://doi.org/10.1007/978-3-030-33043-9_3

identification of potential participants. Scientists can recruit anybody who they believe will be able to contribute to the research objectives and they cannot promise to have any control of whether or how the knowledge created is used by other actors in society. In contrast, participation in decision making must follow democratic principles of representation for participant recruitment and ensure that participation aligns with formal democratic structures. This difference is little discussed in the literature which has often focused on critiquing the lack of public influence and participatory formats that do not deliver the influence promised. In contrast this chapter discusses two examples of participation in environmental decision making that can be considered successful in terms of delivering the public influence intended.

In the following I begin with explaining why and how the notion of 'environmental decision making' is used in this book. Then I outline an organising conceptual framework using the three rationales—instrumental, normative and substantive—identified as institutional drivers for inviting public participation in decision making. Thereafter I introduce and discuss two examples of how environmental participation in decision making can appear in practice. Both examples are from the UK, one concerns the very controversial issue of geological disposal of radioactive waste. The West Cumbria MRWS Partnership was created with the purpose of bringing the local community into a decision of whether to allow the national Government to a search for a potential site for a geological disposal facility (GDF) in the area. The second example focuses on the much more mundane, but perhaps even more challenging, issue of involving local communities with long-term river management. The UK government introduced the Catchment Based Approach (CaBA) which gives civil society non-profit organisations an important role as hosts for Catchment Partnerships. These are tasked with improving water quality and ensure local community engagement with river management. The chapter ends with reflections on the two cases as examples of technologies of elicitation.

Environmental Decision Making and Participating Publics

In this book, the term environmental decision making has been used when many other authors would have talked about governance or management. This is because the focus is on the practical activities in which publics can participate. Decision making is intended to be more descriptive, making decisions about actions impacting in some way on the natural environment is something that institutions and many organisations have to do. Used in this way the term environmental decision making is very open, it does not distinguish between different types of decisions or characterise the actors involved. At the same time 'environmental decision making' as used here is precise, it refers to purposeful actions with explicit aims and objectives. I do not discuss decisions that have environmental impacts as side-effects, but only those explicitly addressing environmental issues. The term environmental decision making partly overlaps with environmental governance, on the one hand and with environmental management, on the other. The notion of environmental governance is used with reference to the complex dynamics of a multitude of actors doing different things, among them decision making. It captures activities from a historical perspective with the complexity added by compromise and negotiation leading to outcomes that are discernible as long-term societal processes.

Environmental governance is a topic of lively academic debate in several disciplines. The mere amount of writing on this topic is an indication of its complexity. The understanding gaining ground in human geography is that historically there has been a shift from top-down government to multi-level, multisited governance (Bulkeley and Mol 2003). Since geographers pay attention to spatial aspects of decision making and implementation, the shift is also one away from a centre that could order implementation. National governments attached to the Decide-Announce-Defend model of environmental policy making have learned that implementation cannot be forced from the centre when local actors are unwilling. This shift is understood to have been brought about through a range of factors—political, economic, social and cultural. It has changed the distribution of agency in relation to the environment, power has become less well-defined, which means that notions of rational decision making, central in engineering and planning, no longer work as they used to. The traditional technical rationality does not extend across the different actors involved. Today there is a wide range of actors operating on different institutional levels and geographical scales, with mandates of different extents. In the context of environmental governance more effort is required to identify the decision making actors, and to understand their respective remits, as well as their relationships. This understanding of environmental governance has implications for public participation, when agency is distributed it is less clear who needs to involve whom in order to effectuate more democratic decision making.

The other notion with which environmental decision making partially overlaps is environmental management. This is a term often used to discuss more concrete actions. Environmental management involves planning actions to address specific, located problems, such as poor water quality in a particular river reach. Decision making in environmental management concerns how to address environmental problems and risks with physical measures, knowledge and public participation. Discussions of environmental management engage natural and social scientists, as well as engineers and often centre on sharing new methods and best practice examples. These discussions are commonly issue based, for example, the extensive discussion about flood risk management, analysing the consequences of policies, the promise of computer modelling methods and ways to engage local stakeholders. Participation is an important topic for discussion in environmental management discourse and sometimes papers employ an STS perspective.

The term environmental decision making is used descriptively in this book, indicating a wide variety of decisions made about the natural environment. This covers on the one hand, concrete and local environmental management decisions, for example, whether to build a flood defence and what type of defence to build. On the other hand, it also covers high-level policy decisions, such as the adoption of a catchment management plan or an international framework. These are very different activities that have in common a need to reach a decision about what to do, and when the decision has been taken political and legal action is often needed to change it. These environmental decisions change things for people in different ways, ranging from the taxes we pay on fossil fuels, to whether we can sit down on the bank of our local river and dip our feet in the water, or if the water is too polluted.

Environmental decision making today recognises the right of affected people to have a say. This was discussed in Chapter 1 as an important origin story for environmental participation. We saw how the crisis of environmental top-down government escalated from the 1960s to 1970s and how environment movements emerging in affluent post-war countries prompted increased public involvement (Chilvers and Kearnes 2016). Environmental movements contributed to putting environmental protection on the political agenda and to include public participation in international environmental policy. The 1992 UN conference on the environment in Rio integrated public involvement in environmental policy and its implementation.

Rationales of Public Participation in Environmental Decision Making

That high-level environmental policy embraces public participation in decision making is one thing, to implement it in local environmental management is, however, a very different issue. There is extensive academic discussion on environmental governance and the role of the public in it. Empirical studies of more or less successful attempts to engage publics have prompted a conceptualisation widely used in STS, that suggests that it is possible to identify three rationales driving institutions to engage publics in environmental decision making—normative, instrumental and substantive.

A normative rationale is understood to rest on principles of 'democratic emancipation, equity, equality and social justice' (Stirling 2007: 220). Democratic ideals have been explicit in environmental policy since the Rio Declaration created at the 1992 UN summit. In the EU the Aarhus declaration explicitly states that the public has a right to participate in decision making and in EU policy this links with the subsidiarity principle about making decisions at the level closest to the problem. Public participation is also included in EU environmental policy, such as the Water Framework Directive and the Floods Directive. From a normative perspective public participation in environmental decision making is in itself a good thing and does not need any further justification. The normative rationale is very explicit in international and national environmental policy and its main challenge is effective implementation—enabling affected people to participate in meaningful ways.

An instrumental rational for public participation in environmental decision making looks for a 'better way to achieve certain ends' already defined by the institutional actors (Stirling 2007: 220). Participation driven by this rationale aims to implement policies, strategies and plans already decided upon by decision making actors with minimal opposition. The idea is to create wide consensus about the choices already made to facilitate effective implementation and create public legitimacy. Participatory activities in which this rationale can be identified have been subject to criticism by social scientists because the ability of participating publics to impact on the decision making is very limited. However, some more in-depth reflection could trouble this as a general conclusion.

In a representative democracy that promotes evidence-based environmental management, it is not self-evidently un-democratic for, e.g. a local authority to decide on and implement a management strategy to address an environmental problem without consulting the public on possible management options. To automatically distrust formal structures of environmental decision making would only make sense if they often or always failed. And, if public participation would be systematically associated with more effective ways to address environmental problems and risks. As long as we support representative democracy as the preferred system for making collective decisions, we have to recognise that an instrumental rationale for public participation makes a lot of sense, it satisfies the rights of people to be informed and to have a say, would they so wish. It may seem unnecessary to state this but, when the reason for studying environmental participation in governance is often to examine occurrences of failure and conflict, very few social scientists bother with explaining that an instrumental rationale can be legitimate and bring about excellent outcomes for a local community.

However, in cases when institutional knowledge and expert advice do not suffice to solve environmental problems participation guided by a substantive rationale has the potential to lead 'to better ends' (Stirling 2007: 220). This rationale promotes the invitation of publics to contribute ideas for how to solve problems and input to decisions about how to move forward with environmental management. The purpose is to widen the range of knowledge and values involved, and practical, experience-based knowledge is considered as equally valuable for finding solutions to complex problems as science and technical expertise. There is a strong case for the potential of environmental participation to prompt innovation underpinning the substantive rationale.

The three rationales capture the drivers for institutions acting on different decision making levels in society to involve publics in environmental decisions. If we consider the perspectives of the participating publics two aspects of environmental decision making that really matters are geographical scale and levels of decision making. All three rationales assume that there are interested and possibly knowledgeable publics willing to participate. For this to be the case there has to be a good match between environmental decision making and local matters of concern. People need to view themselves as being affected by an issue to be interested in participating. This has been discussed in the literature by authors concerned about the difficulty to recruit and retain lay participants in participatory decision making. They note that there is often a gap between the mandate of decision makers and the interests of participants, e.g. decision makers may want input from publics on national strategies for managing future droughts while publics are worried about the impact of water management strategies on local biodiversity.

In the following we examine two practical examples of environmental participation in decision making that relates differently to geographical scale and decision making levels. The West Cumbria MRWS (Managing Radioactive Waste Safely) Partnership was created as a result of a new Government strategy that had the ambition to solve a long-standing national problem by local participation. In the second example we see how a new national framework, created to implement EU policy, involved civil society organisations in ongoing river management. The CaBA installed a new type of organisation called Catchment Partnerships (CP) as a way to effectuate both community participation and good water quality.

Public Participation in Local Decision Making: The West Cumbria MRWS Partnership

I learned about the West Cumbria MRWS Partnership (from here on the 'Partnership') when working in a project on the sociotechnical challenges of geological disposal of radioactive waste.[1] At the time the Partnership had just completed three years of in-depth examination of technoscientific knowledge and investigated local community views on whether the locality should volunteer to take part in a siting exploration for a GDF for radioactive waste.

High-level long-lived radioactive wastes from nuclear power plants and military weapons production are some of the most toxic materials ever created. How to dispose of them poses a challenge that remains unsolved. Since the 1970s there has been international consensus among scientists and policy makers that the safest disposal method for these wastes is deep geological burial. Most countries that have such waste have developed plans for burying it deep underground, where it can lay safely for many hundred thousands of years until it is no longer dangerous. However, implementing these plans has proved much more challenging than anticipated by politicians and experts in the 1970s. Every programme for geological disposal of nuclear wastes has encountered active public opposition.

---

[1] The 2011–2014 InSOTEC (International Socio-Technical Challenges for implementing geological disposal) collaborative social sciences research project, funded under the European Atomic Energy Community's 7th Framework Programme was led by Dr. Anne Bergmans, University of Antwerp.

The public opposition to geological disposal has been highly effective, locations identified as suitable for such experiments have been protected by local publics and allies rallying to their support. In the 1990s governments, whose geological disposal programmes had stalled in the face of public opposition, turned to social scientists to get help to understand why people resisted what was considered the only rational solution to the problem. One outcome of the knowledge about science, technology and society generated was a reformulation of policies in a way that was conceived of, by governments, as increasing public influence and democracy through participation.[2]

Implementing this new, more democratic, approach several European governments developed 'voluntaristic' programmes to find sites for geological disposal. The key idea was that a local community somewhere in the country would volunteer to host a GDF. The locality taking on this task would be compensated financially, get a number of new jobs and become a centre of science and technology innovation and expertise. The UK embarked on such a programme, called Managing Radioactive Waste Safely (MRWS), in 2008. By June 2009 three local authorities had submitted formal 'expressions of interest' to participate in a scoping exercise that involved assessing the geological suitability of the area and establish the views on the issue in the local community. A new organisation was created by the three local authorities to undertake a joint investigation and recommend whether the councils should embark on the next step in the MRWS process, or not. The West Cumbria MRWS Partnership commenced its work in 2009 and delivered its final report in 2012 and it was probably the most extensive local public engagement activity ever undertaken in the UK, with regard to the time used, the number of activities organised and the proportion of the local public involved.[3]

The Partnership comprised representatives appointed by the three local authorities, some were council officers and other were nominated by the councils' stakeholders. All were lay people with regard to the scientific and technical expertise of radioactive waste disposal. The Partnership was

---

[2] For in-depth analysis of this issue see, for example, Bergmans et al. (2015), Blowers (2017), and Chilvers (2007).

[3] Detailed information about the partnership is archived at http://www.westcumbriamrws. org.uk, the work is summarised in the final report: West Cumbria MRWS Partnership. 2012. The final report of the West Cumbria Managing Radioactive Waste Safely Partnership. Copeland Borough Council and 3KQ Ltd. ISBN 978-0-9573709-0-6.

defined in terms of reference agreed upon by the councils and the funding came from the UK government, via the relevant department and the Nuclear Decommissioning Authority (NDA). The Partnership had to submit funding applications, with proposed budgets, for one year at the time.

We studied the Partnership as a historical phenomenon, after it delivered its final report, and elsewhere we have written about the ways in which this public engagement changed the way in which technoscience was done (Landström and Kemp 2019), but the present focus is on the way in which the Partnership constructed environmental participation.

As there was no precedent to the MRWS programme and only one Partnership working with this issue in the UK they had considerable freedom to develop a programme of activities. It would have been possible for the Partnership to consider itself to represent the local community, but they did not. Instead they devised an extensive programme that would allow as many local residents as possible to express their views. This was done in the form of a Public and Stakeholder Engagement (PSE) programme comprising three 'rounds' addressing all aspects of GD. Several expert consultants were commissioned to carry out different tasks, using qualitative and quantitative social science methods, as well as deliberative techniques. The PSE programme was comprehensively recorded, with reports presenting and summarising each stage, each activity and each meeting.

The first round of engagement—PSE1—running from November 2009 to March 2010, had among the key objectives to: 'Seek input from stakeholder organisations on the Partnership's work programme, Terms of Reference, Criteria and PSE Plan' (West Cumbria MRWS Partnership 2010: 5). That this fundamental question was posed sets this engagement process apart from the type of events we identified as following an instrumental rationale. To ask for the local publics' views on the organisation inviting them to participate is very unusual. Potentially the answers could prompt a reorientation of the project as such or reshape its ways of working. Whatever the outcomes, posing this question demonstrates a rationale very different from the instrumental.

While the Partnership arranged participatory activities, they also visited already established local public spaces to give presentations. One example was Neighbourhood Forums (NF). Partnership members attended these evening meetings and gave a standardised slide presentation that explained GD, the role of the Partnership and the PSE programme. Representatives from the institutions and organisations responsible for radioactive waste management in the UK also attended NFs. This contrasts with the literature criticising the organisation of participatory events in isolation from decision making or research.

According to the written accounts over 500 local residents attended the NFs and the detailed minutes quote discussions that show very diverse views on the issue in the locality. In addition to the different views expressed by the local residents attending the events there was also well-known long-time established opponents of GD who did not get involved. Organised critics declined invitations to formally engage with the Partnership, rejecting the reopening of the question of siting a GDF in the locality. From the outset the divided opinions of local residents were made visible by the Partnership.

Sensitivity to the diversity of local publics is also visible in the quantitative attitude surveys conducted by the polling organisation Ipsos MORI, complementing the face-to-face activities. The first survey, carried out in November 2009, covered what in any other context would have been described as a representative sample of the population. But, showing the contentious nature of establishing public opinions on this issue the report explains that only a sample of the residents of the area had been surveyed and that this meant that the local community was not comprehensively represented. The first PSE round brought several issues to the Partnership's attention of particular interest was a need to clarify decision making. To continue to deliberate on the issue in the local community it was critical to make clear who would be mandated to decide about what, and when. Another issue was that people wanted more information about all aspects of the issue.

The Partnership's second PSE round, running from November 2010 to February 2011, focused on public education. Again, a wide range of local activities were organised, for example, 10 one-day events organised by 3KQ, the expert consultant firm commissioned by the Partnership to facilitate the process and support it. Each such event comprised an exhibition that explained the issue and there were technical experts representing the institutions responsible for managing radioactive waste, implementing waste management strategies, scientists from the British Geological Survey and representatives of the regulatory bodies ensuring that all rules were followed, all ready to answer questions. There were also other face-to-face events, organised by various consultants, experts in public participation. At the end of the round another opinion survey by Ipsos MORI showed that awareness of the issue had increased, but also that the confidence in the Partnership and support for locating a GDF in West Cumbria had decreased slightly. As in many cases highlighted by STS more knowledge did not lead to positive appreciation of Government policy.

The third round of PSE was a more traditional type of public consultation. A draft of the Partnership's final report was published and local residents were invited to comment on it. Again, the Partnership went further than many other organisations would do and arranged group deliberations as well as inviting written submissions from individuals. Ipsos MORI did a third survey, adopting the same cautious language as in the first one and they found that there was no community consensus on this issue.

After three years of work the Partnership delivered a report that did not present any recommendations, instead they put forth 'lessons learned'. Viewed from the critical STS perspective of this book this is significant because it shows that it is no easier to achieve consensus through public participation than by any other means of decision making. The national government wanted to create local consensus to make it possible to implement geological disposal. However, it became very clear that more deliberation does not create consensus on topics that are controversial, uncertain and that have a multitude of unknown future implications.

As a technology of elicitation, discussed in Chapter 1 as constitutive for the character of the participatory activity and the identity of participants, the West Cumbria MRWS Partnership can be understood as an experiment in public participation in decision making. It was constituted in a national policy devised by national Government driven by an ambition to create legitimacy and becoming able to implement GD through ensuring that a local community could be seen to volunteer to host a GDF. The Partnership was expected to generate community consensus that would underpin a decision to proceed. Interestingly the Government's expectation of being able to manufacture such consensus intersects with STS critics' contention that this type of participation based in an instrumental rational can only affirm institutional ideology. However, from the perspective of the national Government the experiment failed, no community consensus to go ahead with the process of siting a GDF ensued.

This was an example of a one-off experiment in environmental participation. The Partnership's work was completed before the political decision was taken and regardless of the outcome it would not have been repeated. If it had led to the community deciding to volunteer to take part in a siting process it would probably have become an international exemplar, recreated in other countries with the same problem. Being one-off-events is a common feature of participation in environmental decision making. Quite often institutional decision makers, public as well as private, want public participation in relation to specific choices construed as decisive. Such one-off decision making to solve an issue can appear dated, as belonging to the high modernism of the 1950s and 1960s when all environmental problems, natural and human-made, could be resolved by new, preferably big, engineering projects. In the 1970s this conception of environmental decision making was challenged in ways that brought participation legitimacy, but it did not dismantle the idea of making final decisions that would settle issues. For this to change two other elements have been critical, firstly the shift, driven by wider social processes but also by environmental participation, from top-down government to multilevel, multicentred governance. Today institutional actors do not have mandate or resources to embark on engineering solutions on their own. The management of environmental problems and risks demand collaborative strategies. The second shift has occurred within environmental governance as a reaction to a new scientific understanding that environmental change is inevitable and ongoing.

The idea of making final decisions about how to manage the environment and its risks is based on a notion of the natural environment being stable. The new scientific understanding of environmental change, associated with climate change and more local processes that are directly experienced such as soil erosion due to deforestation, have also contributed to another approach in decision making. Notions like adaptive or anticipatory governance, point to the need for continuous review and revision of environmental management strategies. By some this shift is also seen as an opportunity to democratise environmental decision making, to make it more open and tentative and responsive to experience, practice and local knowledge.

CATCHMENT PARTNERSHIPS—CONTINUING ENVIRONMENTAL PARTICIPATION

Catchment Partnerships are comprised by the heterogeneous population of stakeholders affected by surface water and interested in its management; they explicitly involve local communities, in addition to institutions and businesses. Catchment Partnerships were created as a key feature of the CaBA first introduced in a Defra (Department for Environment, Food & Rural Affairs) policy paper in 2013. The driver for this policy was the recognition by the UK Government that water decision making and action needed better local connection. This system was seen as being very top-down in a way understood to create barriers to the implementation of many measures that would promote better water quality, something required by the EU Water Framework Directive (WFD) adopted into UK legislation. In addition to failing the WFD water quality criteria the UK also had problems with getting local communities engaged with water management and policy. Making the catchment the primary unit in water management forced new collaborations since catchments do not coincide with existing political and administrative units. It also brought local decision making to the forefront in ways that made water management and governance relevant to local people. The 2013 policy paper was underpinned by a pilot project initiated in 2011 and it was the success of this that encouraged the nationwide implementation of the policy.

The 2013 policy paper has six sections explaining the objective of the CaBA, how it would work in reality, the relationship to other Defra delivery schemes, what Defra would do to support the local adoption of CaBA and how the transitional arrangements would look. Key to this document is the definition of a catchment as '[A] geographic area defined naturally by surface water hydrology. Catchments can exist at many scales but within this framework, we have adopted the definition of Management Catchments that the Environment Agency uses for managing availability of water for abstraction as our starting point' (Defra 2013: 3). As indicated in the quote CaBA introduced a new organisational unit, the Catchment Partnership. These were defined as: 'a group that works with key stakeholders to agree and deliver the strategic priorities for the catchment and to support the Environment Agency in developing an appropriate River Basin Management Plan, required under the Water Framework Directive' (Defra 2013: 3).

According to the policy document the Environment Agency (EA), the public agency leading on water issues in England, was tasked with the responsibility of implementing Catchment Partnerships in all river catchments in England. The partnerships themselves were explicitly underpinned by an idea of stakeholder co-management and from the outset civil society organisations were viewed as central. As Catchment Partnerships were set up, local environmental organisations—often in the guise of rivers trusts or wildlife trusts, non-profit, volunteer-based—took on the role of hosts. This positioned these, often quite small, local environmental stewardship groups in a new relationship with powerful actors in water management such as the water utility companies and the EA.

Working with several local rivers trusts in England we learned about how the CaBA strategy was experienced as it was rolled out. Rivers trusts are non-profit charities that organise local volunteers who want to get involved with their local river. People volunteering have different interests and many are attracted by the opportunity of hands-on involvement with environmental improvement. Local rivers trusts organise practical projects, ranging from picking litter in the river and on the banks, to building new wetlands, to water quality monitoring in citizen science projects. Rivers trusts are propelled by an ethos of caring for the river, in a relationship that is not based on ownership or any economic interest but on people's desire to do something meaningful in terms of protecting, restoring and maintaining river environments.

Rivers trusts comprise local residents, there is often quite a lot of retirees since they have the time to become more involved and they can often contribute important skills acquired in their working life. However, many younger people and children become involved with rivers through specific campaigns launched by local rivers trusts, for example, primary school classes breeding fish over a few months and then releasing them into the river. The larger rivers trusts can have one or more project officers employed, often they have the capacity to apply for the project funding required to create more such posts. Rivers trusts with project officers tend to become rather active in developing local river improvement projects in which local businesses are commissioned to do the work. The most successful rivers trusts can become effective delivery organisations for interventions mitigating flood risk or improving water quality.

Involving local environmental stewardship groups, such as rivers trusts, can be seen as a democratisation move that increases local participation in river management and makes sure that local communities can participate in decision making. Rivers trusts collaborate widely with other actors in their localities. They develop relationships with other organisations, public and private, that can assist them with resources and knowledge to turn their vision for local environmental improvements into reality. These relationships can be rather complicated, for example, many of the rivers trusts in the Thames Valley collaborate with the big water utility company Thames Water on a variety of local projects. Hence, they can depend on this corporate actor for much of their funding while also opposing some of the actions undertaken or proposed by Thames Water, such as abstraction of water from rivers or the building of new reservoirs. Equally complicated relationship can be found between local rivers trusts and the EA. At first glance rivers trusts and the EA appear to have converging interests, to protect rivers and their environment. However, the ways in which this is to be achieved and who can do what may be less well aligned. In many cases rivers trusts find that the EA want them to take on tasks that the EA is no longer funded to do, or that they are not really in a position to do, at a much lower cost than if the task had been done by the EA themselves. Viewed in this light CaBA is not purely a way to engage civil society in water management, but it can also be one way of reallocating responsibility from EA, or the public sector, to rivers trusts, civil society actors.

In the first years of CaBA we met people involved with local rivers trusts who were ambivalent about playing a more central role in catchment management. On the one hand these local stewardship groups were often quite small with limited human and financial resources. Hosting a Catchment Partnership could be more than they could manage and threaten to exhaust the volunteers and make them leave. There was also a sense of being marginalised in the CP, by the national institution EA with a long history of building and using expert knowledge in water management and the water utility companies, that had the corporate resources needed to pursue their interests. Disagreeing with these two 'partners' seemed rather pointless to many of the volunteers we met. As catchment hosts, local volunteer groups can feel overrun and outmanoeuvred by the institutional and corporate actors who are also members of the Catchment Partnerships. In social science the situation of the local groups in CPs could be considered to be co-optation. This is the case in situations where activist, or civil society groups, become captives in organisational frameworks in which they have very little capacity to influence the decisions made in their name (cf. Jamison 2001). In this case the CaBA CPs could be seen as a way to silencing potential critics in civil society.

Another challenge that people mentioned was the funding, Local nonprofit groups received the funding needed to act as CP hosts from the EA (who are also members). CP hosts have to apply annually, which can give rise to doubt about the durability of the CaBA framework When we first heard about CPs from people involved with local rivers trusts there was concern about the very limited funding offered to act as partnership hosts. This was a novel task for many small stewardship groups with activities based on the free labour of the volunteers. Hosting a CP would mean a significant commitment to be involved in regular, long-term, activity which would pose a challenge to small volunteer groups. The ambivalence felt among local volunteers about their group taking on the role as CP hosts points to the challenges posed by public participation in decision making via local organisations. In the beginning the EA struggled to find local groups that were willing and able to host CPs.

However, things changed with time and today CaBA appears to have been claimed more by the non-governmental actors. The current website says that: '[T]he Catchment Based Approach (CaBA) is a community-led approach that engages people and groups from across society to help improve our precious water environments. CaBA Partnerships are now actively working in 100+ catchments across England and Wales'.[4] This shows how at least some of the people involved with the participatory aspects of CaBA and CPs have adopted the approach in full. We can guess that as more CPs have come into action the expectations on what local stewardship groups can contribute and how the collaborations work, have evolved and become better aligned with the financial and human resources available, making CP hosting manageable. If this is the case CPs could become an enduring way to integrate public participation in water quality management. It is already viewed as an example of success among scholars in other countries. The UK is often mentioned as leading the way with regard to how the local participation required by the EU WFD in all countries can effectively be encouraged. However, the stability of CaBA can be questioned, the one-year at the time financing arrangement of CPs means that the framework could very rapidly be dismantled if policy objectives would change.

THREE RATIONALES IN PRACTICE

Considering the complexity of the two practical examples of environmental participation in decision making discussed in this chapter there is no surprise that there is extensive academic debate across several disciplines on the many aspects of this phenomenon. To indicate some of the intrigue we can start with the three concepts introduced in Chapter 1 and at the beginning of this chapter—instrumental, substantive and normative rationales. These concepts formed a framework that was very useful to highlight important aspects of the cases but, as all conceptual schemes do, it left many things out.

---

[4] https://catchmentbasedapproach.org/quote, July 30, 2019.

If we look at the West Cumbria MRWS Partnership, we find that all three rationales are present, but that they are expressed in the work of different actors. The UK Government acted with an instrumental rationale they saw participation as a way to achieve their objective of siting a facility for geological disposal for radioactive waste. They also defined the public as being the local community, a notion that became a challenge in practice since it did not correspond with existing formal democratic decision making structures. The elected officials in the local authorities were not seen as the legitimate representatives who would be able to take the decision to volunteer for a siting study without explicit agreement by their constituency. The Partnership was the motor in a technology of elicitation that would affirm the mandate of the local authorities to decide on this particular issue. The local authorities apparently acted with a more substantive rationale if we understand them to want the most comprehensive knowledge about both the technoscientific and the social aspects of the issue. In distinction to the UK government the local authorities did not unite around the goal of implementing geological disposal in the locality. The terms of reference they based the Partnership upon emphasised knowledge and fact-finding. This may be a consequence of the local memory of previous public controversy over geological disposal in the locality. Uncertainty about the material impacts and the sociopolitical viability of geological disposal that fuelled previous conflicts could potentially be reduced through the work of the Partnership. The Partnership itself expresses a normative rationale. The most important thing was to make sure that every local resident was provided opportunity to make their voice heard.

With regard to the three concepts the CaBA example is equally complex. Again, the UK government was clearly acting with an instrumental rationale, they wanted participation in order to implement the WFD requirements physically and politically. In this case it was the national government that invented a new kind of environmental decision making body, the Catchment Partnership. The remit of CPs remains unclear as far as we can see, however, if the actors with formal and economic power act upon the plans developed in the CPs they will have impact. The creation of CaBA was driven by an instrumental rationale but its implementation provides space for both substantive and normative rationales. The actors coming together in CPs all want to contribute to catchment management that will improve environmental status, although they may not necessarily agree on what constitutes the best knowledge and techniques. It is also likely that what is the best solution for a water utility company is not ideal from a local community (for example new reservoirs that will ensure water supply will flood large areas in a locality). The civil society groups acting as CP hosts also tend to find it important to engage local communities in making plans for the catchment while an institutional actor, such as the EA, might consider such groups in themselves as satisfying the need for public representation.

The reality of environmental governance complicates environmental participation in decision making in a way that requires more systematic empirical study. The usefulness of the established concepts is limited when environmental decision making has changed, away from top-down, centralised hierarchy to the distributed, multisited and multilevel reality signified in the term governance.

'Technologies of Elicitation' in Environmental Decision Making

In this chapter we have looked at two very different ways in which local publics have participated in environmental decision making. In both cases the challenge was to find ways to encourage and facilitate local public participation that was demanded in Government policy. The West Cumbria MRWS Partnership was a locally organised 'technology of elicitation' (Lezaun and Soneryd 2007). The close examination of the Partnership showed how much work it took to generate a comprehensive understanding of how a local population considered the controversial issue of GDF siting. In the example of the Catchment Partnerships articulated in the CaBA the challenge was twofold, on the one hand, to improve the quality of surface waters in England and on the other hand, to engage local communities with water management. The community engagement was thought to be necessary to gain support for activities and interventions to improve water quality. This conception of public participation echoes the Rio Declaration. In the first case there was an end point, the public participation was to lead up to and inform a decision that would be made by elected representatives. In the second case there is no end point, the CaBA system will continue as long as it is considered useful by the national government.

Both the Partnership and CPs qualify as 'technologies of elicitation' in the sense that they were created with the purpose of engaging local lay publics and articulate their knowledge and values in relation to the issues. These complex apparatuses involve the creation of ways in which publics can be convinced to take part and express their knowledge and opinions. There are many other such technologies at work to facilitate environmental participation with more or less successful outcomes. The important thing is that they are necessary, environmental participation in decision making requires knowledge, skill and determination to happen. There is no way to simply ask 'the general public' about something. In the case of the Partnership we saw how they commissioned expert consultants to organise events and document the work. This points to the emergence of a new field of professional expertise that we will return to in the concluding chapter.

Paying attention to the technologies of elicitation needed to effectuate environmental participation in decision making also highlights the mediation of the relationship between decision makers and participants. This mediation is critical for the way in which we can see the three rationales expressed in the examples of participation discussed above. Although very different the Partnership and CaBA were both instigated by national government trying a new approach to problems (radioactive waste and water quality) in line with an instrumental rationale for participation. Both connect with international policies and agreements that the UK is committed to follow and tries to implement, these concern both physical processes and an obligation to involve stakeholders and publics. These similarities result from the conditions of environmental decision making in the context of governance. National government is the top-level decision making authority, but it operates in a space structured by transnational and subnational legal and political relationships.

In order to implement environmental policies, local publics have to be engaged, and this is where intermediaries enter the process. In West Cumbria the Partnership, jointly constituted by the local elected authorities, made it their aim to engage with as many local people—individuals, groups and organisations—as possible in order to capture their views on the issue. This is a different rationale, and in the documentation left behind we can discern a normative rationale in the Partnerships work. They considered it important that everybody got to exercise their right to have a say. This undertaking fitted perfectly with a notion of democratisation however, it did not lead to the delivery of a clear answer. This points to the fact that any community is too diverse to generate clear for or against on decisions involving significant uncertainty. Taking the diversity of views in any community seriously participation cannot be the answer to what to decide. The failure of instrumentally pursued participation is inevitable in contemporary society. Either a technology of elicitation must be devised to manufacture consent, which can create serious controversy. Or the apparatus really captures the genuine views of publics and will thus represent diversity and disagreement. In relation to radioactive waste the first was the strategy that led to failure in the 1990s and the second was the outcome of the Partnership.

In the case of CaBA decisions are oriented towards the level of local physical management, concerning how to intervene in catchments to improve water quality and reduce other risks as well. The civil society groups involved in the CPs are positioned as representatives for local communities, however, we know that they are not demographically representative. Environmental stewardship groups tend to be dominated by middle aged, middle class, educated people from the dominant ethnic group in a locality. This is not really a problem as long as the group is advocating on a particular issue or organising events to get people to spend time by the river. However, when these groups become involved in decision making it can amount to delegation of decision making power from elected authorities (or the agencies working on their behalf) to self-selected advocates for the interests of the privileged. This may not have led to open conflict yet, but it certainly has the potential. The new role of local environmental groups in participatory decision making highlights the character of environmental governance in which decision making agency is multisited and multilevel.

References

Bergmans, Anne, Göran Sundqvist, Dragos Kos, and Peter Simmons. 2015. The participatory turn in radioactive waste management: Deliberation and the social-technical divide. *Journal of Risk Research* 18: 347–363.

Blowers, Andrew. 2017. *The legacy of nuclear power*. London: Routledge.

Bulkeley, Harriet, and Arthur P.J. Mol. 2003. Participation and environmental governance: Consensus, ambivalence and debate. *Environmental Values* 2: 143–154.

Chilvers, Jason. 2007. Democratizing science in the UK: The case of radioactive waste management. In *Science and citizens: Globalization and the challenge of engagement*, ed. I. Scoones and B. Wynne, 237–243. London: Zed Books.

Chilvers, Jason, and Matthew Kearnes. 2016. Science, democracy and emergent publics. In *Remaking participation: Science, environment and emergent publics*, ed. J. Chilvers and M. Kearnes, 1–27. London and New York: Routledge.

Defra. 2013. Catchment Based Approach: Improving the quality of our water environment. A policy framework to encourage the wider adoption of an integrated Catchment Based Approach to improving the quality of our water environment.

Jamison, Andrew. 2001. *The making of green knowledge: Environmental politics and cultural transformation*. Cambridge: Cambridge University Press.

Landström, Catharina, and Stewart Kemp. 2019. The power of place: How local engagement with geological disposal of radioactive waste re-situated technoscience and re-assembled the public. *Science and Technology Studies*, online first.

Lezaun, Javier, and Linda Soneryd. 2007. Consulting citizens: Technologies of elicitation and the mobility of publics. *Public Understanding of Science* 16: 279–297.

Stirling, Andy. 2007. Opening up or closing down? Analysis, participation and power in the social appraisal of technology. In *Science and citizens: Globalization and the challenge of engagement*, ed. M. Leach, I. Scoones, and B. Wynne, 218–231. London and New York: Zed Books.

West Cumbria MRWS Partnership. 2010. Public and stakeholder engagement round 1 report. Adopted 13th May (Doc. 61 in the electronic archive at http://www.westcumbriamrws.org.uk).

CHAPTER 4

Public Participation That Reconfigures Expertise

Abstract First, I explain how expertise can be understood as bridging the gap between science and decision making. Thereafter the notions of uninvited and invited publics are introduced as a conceptual starting point for discussing the relationship of environmental participation and expertise. Starting with uninvited public participation the first practical example is of environmental justice. Then we turn to environmental participation in expertise by environmental movements and discuss the involvement of scientists who are critical of expert knowledge claims. Finally, I discuss community modelling, an example of how academic science can assist local environmental groups with knowledge and tools to increase their capacity to participate in expert-led decision making.

Keywords Expertise · Environmental justice · Movement expertise · Community modelling

Expertise Bridging the Gap

It may appear strange to introduce expertise as a way to bridge the gap between science and decision making when this distinction was introduced and justified in Chapter 1. However, the point was not to argue that there is no connection between the two but to be able to specify the link in more detail. The theoretical conception of the entwinement of science and

C. Landström, *Environmental Participation*,
https://doi.org/10.1007/978-3-030-33043-9_4

governance established in STS is a critique of a philosophical view stating that science is autonomous and not affected by politics or values of society. Critiquing this idea empirical studies of science as social practice demonstrate the close and mutually supporting relationship between science and the society in which it is pursued. The ambition has not been to argue that science and decision making are the same activity. As the entwinement of science and governance as institutional processes has become established as common sense knowledge in STS public participation is often discussed as a point of intersection for science and decision making without much precision. In contrast, I insist that more precision as to how this intersection works is needed to understand environmental participation in practice.

The separation of science and decision making made in this book is pragmatic, based in the study of how publics participate in different practical activities. On the ground, in local practices there is a difference between participating in science and participating in decision making. Conflating the two can lead to failure to, for example, retain participants. This has been the experience of scientists who have invited publics to engage with knowledge production that turns out not to have the envisioned connection to decision making. Participatory decision making can also result in disappointment and conflict if it, in the end, resorts to expert-led management, treating involved publics as less important. Identifying expertise as a concrete link between science and decision making makes it possible to specify how the gap between the two is often overcome in practice. For this to be illuminating we need to clarify the notion of expertise.

In a discussion of expertise in the context of public discontent and the growing heterogeneity of knowledge production Helga Nowotny notes that 'the demands for scientific and technical expertise in a socially distributed knowledge production system differ from what research normally can provide' (Nowotny 2000: 15). Illuminating the historical dynamics of expertise Nowotny explains that the reason for it to become the site where science and policy meets is partly due to an ideology that separates science from society. This separation creates a need for a way to bring scientific knowledge to bear on societal questions and problems without science being culpable for mistakes. Expertise has become science's 'link to action and the possible recommendations for policy recommendations frame the questions as well as the answers' (Nowotny 2000: 15). The enactment of this cultural order is what we discern in the practical separation of scientific research and institutional decision making in environmental participation practices.

While Nowotny focuses on the historical development of the relationships between the institutions of science, expertise and policy this book attends to local practices. From this perspective the difference between science and expertise is not primarily narrative or historical, but practical. Scientists working in universities can pursue their own interest if they manage to find research funding, while, for example, technical experts deliver answers to questions of concern to other actors. For advice on what to do in order to address specific environmental problems decision makers turn to experts, who accept commissions and agree to deliver according to the client's brief. These experts draw on established scientific and engineering knowledge that they apply to local circumstances, in order to produce the advice asked for. Such advice can comprise a number of options that are scientifically sound, technically feasible and that fulfils the cost-benefit criteria acceptable to the decision makers.

One area in which the distinction between science and expertise is quite distinct is UK flood risk management. In a case study we found that the knowledge-base, used by local and regional decision makers, was generated by consultancies with most of their expertise in computer simulation modelling (Landström et al. 2011). When the consultants were commissioned to investigate flood risk and mitigation options for a locality, they selected a well-known and trusted modelling approach that they used to produce a study that presented different measures for flood risk mitigation to local decision makers. The experts working in consultancies were very clear on not being interested in the models as such or in producing general scientific knowledge, they wanted to help people solve problems. In contrast scientists studying flooding were creating new and better computer simulation models that they hoped would be used widely to analyse the causes and dynamics of flooding and flood risk.[1]

The understanding of expertise as bridging the gap between science and decision making makes it relevant for environmental participation. Because expertise is visibly advising decision making it becomes a site of uninvited participation, differently from science. Participation in both environmental science and decision making is by definition invited. Environmental scientists very rarely encounter opposition, the insistence of following their own

---

[1] Whatmore and Landström (2010) show that new scientific knowledge is not necessarily useful to experts.

research agenda and working in institutions formally separate from decision making, makes it rather uninteresting to start with the science if you want to change environmental decisions. For example, to stop a water quality research project in a university will not have any impact on a polluting industry. In contrast, challenging the expertise underpinning the decision to grant the industrial facility a licence to operate in the locality is more meaningful, particularly if combined with public protest.

Different from environmental science, environmental decision making frequently encounter public opposition. However, opposition and protest are not considered to be environmental participation in this book. This is a choice made in order not to de-politicise active resistance. If all public interaction with environmental decision making is viewed as participation the difference between chaining oneself to a railing in order to stop environmental destruction and sitting down in a room to deliberate is obscured. Even if they would have the same impact on, e.g. hydraulic fracturing in a sensitive chalk catchment, the practices differ significantly. Invited participation operates within the existing social order while protesters actively challenge the distribution of power. Importantly, protesters subject themselves to risks that are not present when participating in, for example, deliberative events.

The conceptual distinction of uninvited and invited participation introduced by Brian Wynne focuses on the politics of knowledge. As he explains '[D]eliberately or not, invited public involvement nearly always imposes a frame which already implicitly imposes normative commitments—an implicit politics—as to what is salient and what is not salient, and thus what kinds of knowledge are salient and not salient' (Wynne 2007: 107). What happens when people react to technocratic decision making and band together to challenge it is seen in that '[U]ninvited forms of public engagement are usually about challenging just these unacknowledged normativities' (ibid.). As has also been noted in the previous it is rarely scientific knowledge as such that is challenged through uninvited participation, Wynne clarifies that it is the extension and combination of science with other features that prompts discontent. It is this reconfiguration and extension of scientific knowledge that we can understand as expertise, in line with Nowotny's analysis.

The environmental competency group staged in Pickering discussed in Chapter 2 originated in a context of public challenge to expert knowledge claims. The controversy was partly propelled by local residents questioning the correctness of the conclusion presented by the technical consultants. In that case the scientific project changed the local politics of knowledge, but university research cannot normally be relied upon to provide alternative knowledge claims that critics of expert advice can rely upon. Mostly publics criticising expertise have to be self-reliant, and below we will look at examples of such practices in the US environmental justice movement. When publics choose to engage in criticism of local decision making by creating counter-expertise the term uninvited environmental participation seems appropriate. This is an environmental participation practice with publics developing strategies that allow them to have a say in the decision making by presenting knowledge claims competing with institutional expertise, not by political protest or by rejecting scientific knowledge.

There are more to environmental participation in expertise than being uninvited or invited. There are, as Nowotny pointed out, many sites for production of knowledge. One of these is environmental movements. Some of the organisations emerging in the environmental movements of the 1960s and 1970s are still active. The most successful of these groups have built up science-based expertise that can provide advice with a different perspective from that of corporate or technical experts. This is a different organisation of the relationship of experts and publics, where experts work for volunteer, non-profit organisations.

We also find that publics get invited to participate in production of knowledge that can support local decision making. However, it is not always easy for local publics to accept the invitation and make a meaningful contribution. This is something that science can help them with, drawing on their ability to refashion the tools used by technical experts. After looking at environmental justice and environmental movement expertise we explore one example of how local publics can be supported to work with technical experts and decision makers.

Uninvited Environmental Participation

As noted above one way that publics can participate in expert-dominated decision making processes is through confrontation. Examples of success through such uninvited participation can be found in discussions of environmental justice (EJ). Many cases of this type of uninvited participation have been studied and detailed examination has clarified how such activities have been prompted by the particulars of the US environmental decision making system.

A widely known example of uninvited public participation that impacted on decision making by creating alternative science-based information is the 'Louisiana bucket brigade' studied by Gwen Ottinger (2010). This involved local residents, mostly African American, in Norco, Louisiana, who did not trust the experts declaring that the toxic chemicals released by the adjacent Shell Chemical plant did not pose any health risks. The residents insisted that the measurements generated by the technical experts were flawed, constructed in a way that did not provide a true assessment of the local pollution, ignoring spatio-temporal distribution. The expert monitoring generated average values to satisfy regulatory guidelines. To generate facts that could withstand scrutiny in court the residents recruited scientists who were also critical of the expert claims. With the help of the scientists the residents devised a new kind of instrument, involving buckets and sponges, these sensors were placed where local people perceived pollution as very bad, at the times it occurred. The network of bucket monitoring stations produced data that showed a different pattern of pollution than what company experts had proposed. Through court proceedings the counter-monitoring led to changes in what the company was allowed to release.

The Louisiana Bucket Brigade case is spectacular in that the local public was able to generate counter-science that could provide robust data. Key to this was the scientists that joined to help the residents. In the present context the most interesting thing with this case is that it demonstrates the gap between science and expertise that Nowotny identified as critical. Experts draw on scientific knowledge, but scientists will not necessarily think that they come to correct conclusions in their applications. There is no straight line from generalised scientific knowledge to the expert interpretations of what is going on in a particular place. This is the case regardless of who applies scientific knowledge, contextual factors and choices about how science is applied makes a major difference. The move from scientific knowledge to expert assessments of local processes is based on a multitude of interpretations, experience, rules of thumb, analogies and the like, that professional experts cultivate, but that can always be challenged from another perspective. In the case of the Louisiana Bucket Brigade the way of measuring air pollution that the experts used was premised on their brief from the industry to demonstrate compliance with regulatory frameworks, they did not ask if there were any other way to perceive of the problem. The knowledge generated was successfully challenged by the residents who experienced the polluted air in the place where they lived.

Another example of differently constituted expertise from the United States comes for the environmental organisation American Rivers. Founded in 1973 this national organisation dedicated to environmental conservation brought attention to environmental problems that were not on the scientific or decision making agenda at the time. Campaigning brought volunteers and funding and today the organisation presents itself as the leading river restoration organisation in the United States. This charity counts 355, 000 members nationwide, contributing money and time. Like the UK Rivers Trusts discussed in previous chapters American Rivers combines 'national advocacy with field work in key river basins to deliver the greatest impact. We're practical problem solvers with positions informed by science. With our expertise and outreach, we work to protect and restore our nation's rivers' (www.americanrivers.org).

The American Rivers website highlights the practical work that the charity does around different rivers in the United States but there is also a link to reports and publications that provide a glimpse of the ways in which the organisation produces expertise. The reports published by American Rivers draw on natural science, law and economics, they are funded by named benefactors, sometimes in the form of fellowships for the author. They bring scientific knowledge to bear on problems that the organisation identifies as urgent. One example is a report titled: 'Protecting drinking water in the Great Lakes' that presents the federal 'Safe Drinking Water Act' in relation to a specific context. Another example is: 'Rivers & Roads', a report that explains the problems of stormwater runoff, evaluates ways to manage it and presents two case studies and recommendations. Differently from expert reports commissioned by industry and many decision making institutions, the reports published by American Rivers are written in a way that would make them possible for most of the members to understand.

American Rivers is one of several environmental organisations with roots in the 1970s, bringing attention to environmental problems and insisting on the importance of civil society in addressing these problems. There were no expert-led environmental decision making that they could be invited to participate in at the time. This ties in with the origin story mentioned in Chapter 1 that points to the dual crisis of science and government that became visible in the 1960s and 1970s, when environmentalism became the focus of some of the many social movements emerging in the affluent post-war countries. We noted that the roots of participatory discourse could be traced to the radical science movement that grew in relationships with anti-nuclear protests, anti-colonialist movements and protests against the

Vietnam War (Quet 2014). We look at American Rivers as an example of environmental participation in expertise rather than the Rivers Trusts in the UK, who featured in previous chapters, because of the long history and the contrast with environmental justice also pursued in the United States.

American Rivers is of interest in the context of uninvited expertise because of its social movement roots that are contemporary with environmental science and policy. This organisation has developed expertise constituted in social movement ethics and politics of environmental protection, preservation, conservation and restoration. The scientists working with such organisations investigate questions prompted by a concern for environmental protection, these are not the same questions that international scientific discourses focus on. The experts developing local environmental protection strategies strive to apply scientific knowledge to maximise the health of ecosystems, not as is often the case, minimise the damage casued by resource extraction.

The type of long-term, established organisation exemplified by American Rivers impacts on decision making and the physical environment by grounding volunteer interventions and political advocacy in counter-expertise that draws on scientific research recognised as robust by the international scientific community.[2] The endurance and stability of these organisations provide foundations for environmental participation in expert advice to decision making on different levels in democratic systems.

Another example from the United States demonstrates collaboration between civil society activism and university research. Currently a topic of controversy, and much discussed in the academic literature, hydraulic fracturing (fracking) is subject of much local concern. This new technique for extracting underground gas resources is considered safe by authorities, and the extracting businesses' expert studies of environmental impacts do not find any long-term negative effects. In contrast there are a multitude of non-establishment voices urging caution, if not outright rejecting this activity.

One such concerned local public use various techniques to document local impacts of hydraulic fracturing in a way that is systematised by critical scientist as one activity in the long-running programme 'ALLARM' the Alliance for Aquatic Resource Monitoring. It is led by scientists from Dickinson College in Pennsylvania who started a citizen science project by creating a 'volunteer-friendly protocol in 2010 for early detection and

---

[2] Jamison (2006) discusses how environmental movements have impacted on science as well as decision making.

reporting of surface water contamination by shale gas extraction activities in small streams' (Wilderman and Monnismith 2016: 1).[3] Over time the science-volunteer collaboration has evolved into co-creating analyses and models.

The point of ALLARM's shale gas work is to monitor streams that are too small to show in official and corporate data sets. The thousands of volunteers taking part use scientific measurement kits and also take photographs, documenting what the environment looks like while fracking is taking place. Standard data sets do not show adverse effects of fracking on water ways but close inspection and photographic documentation by volunteers bear witness about negative impacts that have more to do with the physical impact of the activities than subsurface contamination. Small streams and their habitats are negatively impacted by accidental spills and other events that do not register on official monitoring but may have detrimental impacts on local water quality and biodiversity.

As an example of environmental participation this is a practice in which a scientific organisation is in the lead prompted by civil society concerns about an activity that the authorities and corporate sector insist is safe. The counter-monitoring is not set up to challenge the measurement values or technologies used by institutional actors and private businesses, but to capture aspects that they do not have the capacity to observe. This is important since it points to the partial nature of all expertise. Expert knowledge about environmental processes in a place can never be comprehensive and exhaustive. It is always possible to approach issues differently, based on equally valid scientific knowledge. It is the critical stance of ALLARM's monitoring that justify discussing it here rather than as citizen science in Chapter 2. This is science used to critique expertise, showing that it is not primarily science that is challenged by counter-monitoring, but the science-based expertise claiming the environmental safety of all US shale gas extraction.

The examples discussed in this section have complicated the notion of uninvited participation. We started with the obvious case of local activists challenging institutional expertise that supported decision making with negative local impacts. In the United States the notion of environmental justice brings such activities together, as a visible critical social movement involving local activism. I view EJ activism as environmental participation because they challenge expert-based decision making by forming alliances

---

[3] The shale gas monitoring is the most recent ALLARM initiative, the organisation running it has engaged local volunteers with water monitoring for decades.

with scientists who are critical of the specific expert claims supporting the powers that be. The scientists contribute knowledge and skills constitutive to the counter-expertise that EJ action groups use to force changes in decision making.

From a clear-cut example of uninvited participation, the Louisiana Bucket Brigade, I turned to long-term environmental movements that engage volunteers and contribute expertise. These organisations promote goals of environmental protection and restoration, trying to make their science-based expertise more influential than the expertise of extractive, or polluting industry. This environmental participation is neither uninvited, nor invited, it exists in a different conceptual space. Another challenge to the uninvited–invited binary is posed by the example of environmental science supporting grass roots environmental monitoring. In this case, the same type of activity as in local citizen science promotes a critical agenda through monitoring local environmental quality. Distrusting the decision making institutions and the science-based expertise they trust science and scientific methods that they use to build data bases and knowledge about negative local environmental impacts of resource extraction. Again, neither uninvited, nor invited.

The examples of practices in this section have made clear that expertise involving environmental participation can relate in different ways to institutional expertise. We found three such relationships—the direct challenge of EJ counter-expertise, the independent expertise of environmental movements and the complementary expertise of local monitoring. In the next section we look closer at the other concept in the binary—invited participation in environmental expertise.

ENABLING PUBLICS TO ACCEPT INVITATIONS TO PARTICIPATE

In STS discussions invited participation has mainly been treated as the subjection of publics to institutional control. Turning to other academic fields there has been more discussion about the challenges local publics encounter when trying to accept such invitations. Research shows that there are several barriers that prevent publics to accept invitations to participate in the creation of expert knowledge about issues that affect them. One such barrier has been hinted at in the previous—the reliance on science-based technical expertise among institutions and industry. The activities to which publics are invited are often conducted in a language comprehensible only to those sharing the scientific and technical background of the experts. This has been the case in some of the Catchment Partnerships (CP) created through the Catchment Based Approach (CaBA) in the UK, introduced in Chapter 3.

Promoting ongoing, long-term participation of local communities in the management of river catchments local environmental groups are invited to host CPs, thereby becoming central in the strategy. Despite this role these organised publics often find themselves marginalised in the deliberations. The expert knowledge required to impact on the decision making is expensive and primarily available to the water industry and the Environment Agency (EA) who can rely on in-house technical teams or commission technical consultants. This disadvantages local stewardship groups who are experts on connecting local people with water and on organising practical, hands-on activities in the river environments. Such groups do not have access to the science-based technical expertise used, for example, to assess water quality management strategies.

It was the desire to address the unequal access to science-based expertise that prompted the creation of the Community Modelling (CM) approach. This is a technique using participatory methods to enhance the capacity of local environmental groups to participate in expert-informed environmental management, such as CPs. It is intended for use in any area of environmental management in which computer models play a role and so far we have tried it in the context of water management.

The CM format is underpinned by four principles. Firstly, a CM project should be short, no more than six to eight months, including preparation and reporting time. Of this time the participatory activities comprise three to four sessions over a three to four-month period. The relatively short time allocated to a project makes it easier for local participants to attend all sessions. Secondly, resource minimalism is critical for making it possible for local communities to cover the costs of a CM project. A third principle is strategic participant recruitment. The aim is to recruit participants who have the potential to involve other local residents and to continue to engage with the local environmental management process after the project has finished. Fourth, CM uses standardised modelling software. It is critical to use off-the-shelf computer modelling software in free-to-use versions to make sure that the local community can take possession of the computer model used to represent and explore the local environmental issue at the end of a project. Using standard modelling software is also important to make project outcomes transparent to technical experts.

With regard to the purpose of modelling together with local publics it is important to note that the goal is not to make the lay participants able to use the model in the same way as a scientific modeller would. The aim is to let the local participants develop some understanding of how modelling works, how the environmental processes they experience in real life are represented in models and how model outputs are used to consider possible local problem mitigation options.

A key consideration in CM is to ensure that the local people in the modelling group are connected with the relevant decision makers. Depending on the local situation this may already be the case, for example, if CM participants are members of a volunteer group hosting a CP, in other cases this will be a task for the academics running the project.

In the following two different Community Modelling projects in UK localities illustrate the flexibility of these principles in actual local circumstances.

Volunteers Modelling Water Quality in Salmons Brook

The first Community Modelling project took place in London. Scientists from Oxford University worked together with staff and volunteers from Thames 21 (a charity committed to improving waterways and their value for communities) to set up a water quality model for analysis of the water in Salmons Brook, a tributary to the River Lea in East London. The project was initiated by the scientists who wanted to trial this way of working with local publics after having been involved with public participation in science. After a preliminary conversation clarifying the idea, we could arrange a first CM session in Thames 21's office in central London.

Attending this session were two academics from University of Oxford, one social scientist and one natural science modeller who met with four Thames 21 staff members. To begin with, the modeller explained the capacity of the INCA (Integrated Catchment Analysis) model to represent and simulate water quality different rivers. After some deliberation we agreed that it would be worthwhile to apply the INCA model to Salmons Brook where Thames 21 volunteers had built a new wetland with the purpose of reducing the level of pollution in the river. With the wetland completed the volunteers were taking water samples that were sent to a lab for analysis, the measured values of different chemical substances were then noted in a list, but no further analysis of them was done. The group exploring the INCA model thought it would be interesting to use the measurement data in simulations to consider the effects of the wetland and possible future developments. For Thames 21 as an organisation getting some sense of the impact of the wetland was important as they were interested in building more wetlands. A scientific method allowing them to assess the potential effects of such interventions would be helpful in planning new interventions and to argue for their value in the Catchment Partnership that they hosted. It was also something that the volunteers could be interested in. We decided to organise a hands-on modelling session about a month later.

After the first session the social scientist took care of the logistics, booked a venue in London and procured laptops and a portable mini-projector from the university to use in the second scheduled session. The Thames 21 officer who had worked with the volunteers on the wetland invited them to the session and sent the collected measurement data to the modeller. The modeller set up the INCA model to represent Salmons Brook and the River Lea, he also cleaned up and organised the volunteers' data for input in the model.

On the evening of the event about 15 Thames 21 volunteers turned up. Upon arrival everybody was offered tea or coffee and a biscuit, while enjoying the refreshments several people came up to me and said that they were a bit nervous about not understanding the modelling. When everybody had arrived, we started with the scientific modeller presenting the model using the projector. As he walked us through what the model showed and how we could input data (including that collected by the people in the room) and run the model with different river flows people relaxed and got really interested. The visual interface of the model made it possible to see Salmons Brook, a river that all the volunteers knew in real life, on the screen, and the model could represent a process that they all recognised (water flowing through the wetland). We could run simulations showing the measured water quality above and below the wetland using volunteers' monitoring data and then simulate possible responses to questions about what could happen with the wetlands' efficacy if there was a drought or extremely heavy rainfall. By this point nobody was nervous about the modelling any longer.

After the demonstration the people in the room split into three groups with one laptop each to try their hand at running the model. Before the meeting the modeller had set up the model to run on two laptops each handled by a Thames 21 staff member working with a group of volunteers. The scientific modeller guided, supported and explained the activity so that each group could develop a sense for how the modelling worked. The evening went very quickly and eventually the people running the venue came by the room and told us to finish so that they could lock up and leave.

After the second session we agreed to do a third session with Thames 21 staff who would become the custodians of the model and any interested volunteers from the second session. This took place in the same offices as the first meeting, four volunteers showed up, two Thames 21 staff, the two academics and their boss. Again we brought laptops for people to use, but we had also asked participants to bring their own computers so that they would be able to use the model on a familiar machine and keep it as long as they wanted. In this session the modeller went through the process step by step and the social scientist video recorded each step to make a bespoke user's guide that was then made publicly available on YouTube. At the end of the session we officially handed over the model to Thames 21 on a USB stick.

As Thames 21 was already in contact with local decision makers in their role as CP hosts we academics had not arranged any other interaction between them, there was no need as there was an already well functioning relationship. After the completion of this very successful proof of concept exercise in Community Modelling we did several presentations to colleagues in different contexts. We also made a web page with some video recorded follow up interviews with participants. However, the follow-on activities of Thames 21 were much more interesting.

Thames 21 told us that they had gained in confidence and become much better at articulating their ambitions and ideas in the CP. They confirmed to us that the purpose of sharing basic modelling knowledge wider than the closed circuit of expertise involving institutions, corporate actors and technical consultants, had been successful. A completely unanticipated outcome was that Thames 21 applied for funding to create a position for a person who would lead on more community modelling projects with their volunteers using the INCA model in other places in London. The funding they won for this was used to commission the modellers involved with Salmons Brook project to set up INCA models for several other small rivers in London that could be used by local volunteer groups to consider interventions that could improve water quality. This work is understood to have been very useful with regard to Thames 21 planning of water quality interventions where they will be most effective and volunteers' ability to follow up on their monitoring activities.

Affected Residents Modelling Flooding in Otley

The second Community Modelling project was very different from the first. We were contacted by a Town Councillor from Otley in North Yorkshire who was very concerned about mitigating flood risk in the centre of town. Thinking that CM could be a way to enable local residents to engage meaningfully with the flood risk management process we accepted the invitation.[4]

To explore the possibility of co-creating a project with people in Otley we went there to meet with involved town council members to learn about the wider decision making context and about the relationships between residents affected by flooding and the flood risk management authorities. From the town councillors we learned that the authorities responsible for flood risk in Otley were Leeds City Council and the Environment Agency (EA). Otley Town Council has no formal power in this system, but they are consulted, and they have a councillor who attend flood risk management meetings to keep the council informed. The town councillor contacting us was also very engaged and kept in contact with residents who had been affected by the latest flood event and who were at risk of flooding in the future. To us academics it looked like there could be a lot to gain by a Community Modelling exercise that would help the local residents to examine the causes and possible mitigation measures for local flooding in a computer simulation model and get the opportunity to present their understanding of the issue to the decision makers. To be able to undertake a CM project in Otley the social scientists at the University of Oxford and Otley Town Council together applied for and won a knowledge exchange grant. The funds were used to hire two researchers who were going to run the project in Otley.

Employing two researchers to run the project in Otley was an important step in the development of the CM approach. It would allow us to find out whether we could communicate the intent of the activity and the ways of working with the local community sufficiently well, to other academics, to replicate the process from the water quality modelling experience in London without doing it ourselves.

Before the start of the CM activity we arranged a meeting with Otley town councillors, flood risk managers from Leeds City Council, EA, technical consultants and the two researchers in the project. This meeting decided to form an Advisory Group that would receive regular updates

---

[4] For an academic account of this project see Landström et al. (2019).

on the progress of the modelling group from the social scientist and to hold another meeting at the end of the project to discuss the issue with the local participants.

When the research positions had been filled the first step was to explain the idea of strategic recruitment to the social science researcher who was to lead the recruitment of local participants. In this case we did not have the benefit of a volunteer organisation with an already involved membership. Instead the process was more similar to the competency group recruitment that involves interviewing potential participants. Drawing on Otley town councillors' knowledge of who, amongst the residents in the most affected area, had shown interest in the issue the recruitment could begin with a list of people. There were thirteen names on the initial list, all were contacted, and we also asked interviewees to suggest more potential participants. We also contacted the Yorkshire Dales Rivers Trust to inquire about their potential interest in participating. This led to ten interviews with potential participants and of them seven decided to attend the CM sessions.

Since we made direct contact with concerned residents, we took the opportunity to talk to them about how they understood the flood problem, what they considered as the main causes of flooding in the town centre and if they had any suggestion for how it could be addressed. This information was used to make a mind map, representing a wealth of different ideas. This visual collation of individual thinking was brought to the first CM session, both to open discussion and to save time as people did not have to present all of their ideas in the session.

Simultaneously with the recruitment of participants and a social science overview of local views on flooding the natural science modeller took inventory of previous flood risk studies in the area. He looked into historical interventions in the river, what instrument generated data was available and what other flood risk research was taking place in the catchment. Most of what he found was expected, the same types of data that are routinely generated, e.g. rainfall, elevation, river flow, and made available for research use on databases. However, he also found a previous local study with very useful information and he learned about a major flood research programme that had just started. Upon making contact with this programme he was invited to a meeting held to develop a shared understanding of the different flood modelling projects going on in the catchment. The experts working with that programme were very interested in the local participatory activity in Otley and there was agreement on exchanging information between the two initiatives.

With the preparations complete the first session of the CM examining flooding in Otley could commence. When they met for the first time the discussions led up to collective decisions about what to represent in the model and what questions to address. On the scientific modellers advise they chose a well-known hydraulic, free-to-use, software package called HEC-RAS.

HEC-RAS is a computer programme for one-dimensional representation of flow in a channel, i.e. water flowing down a river. It was created by the US Army Corps of Engineers and it can be used to simulate flooding when the water volume becomes too great to be contained in the channel. The HEC-RAS software is free to download and use, but it is complicated to set up and a lot of effort was required by the natural science modeller to make it useable in the group. However, HEC-RAS was approved for use by technical consultants contracted to provide expert assessments in local flood management in England in a benchmarking exercise in the early 2000s. It is still widely used for modelling river flooding in small areas and technical experts are familiar with this software.

After the first session the modeller set up HEC-RAS to represent the reach of the River Wharfe that run through the site at risk of flooding in Otley. He used elevation data and other surveys of the topography to capture the landscape features that affects inundation patterns during flooding. The area of concern is small, and the model was set up with a very fine resolution (50 cm) to capture in the detailed layout of roads, paths, ditches and drains, houses, and so on.

To the second session the modeller brought a basic set-up of the software (not yet representing local river flooding). In the session the group members used the software on laptops to draw their own river channel transects and familiarise themselves with the layout. The social science researcher wrote reports of each session that were shared among group members and sent to the Advisory Group. A local participant posted information about the group and its activities on a Facebook page dedicated to discussing flooding in Otley.

In preparation for the third session the modeller had checked the quality of the local model representation and prepared some flood scenarios. In the session the group members were able to explore flooding in Otley with the model set up as to represent the most flood-prone area in town with different flows and the likely inundation zones, depths and velocities. The local group members who had experienced the most recent flood event were surprised and impressed by how well the modelled flooding agreed with their memories and recordings of the real thing.

In the final fourth session the modeller led the group in a summary of their understanding of local flooding and they explored possible management options. The social scientist contributed to the collective creation of guidance materials that would enable the local participants to use the model and run scenarios after the completion of the CM project.

After the final modelling session, the two project leads from Oxford organised an Advisory Group meeting with local participants and flood risk management decision makers from Leeds City Council and the EA. In this meeting we found that the modelling activity had provided the local group participants with a new vocabulary that enabled them to articulate their understanding of and concerns about local flood risk in a way that the decision makers could understand.

Rethinking Environmental Participation in Expertise

Looking back to the concepts used in Chapter 2 discussing environmental participation in science, the definition of 'education' as a one-way relationship in which scientists teach publics, we could view CM as simply an exercise in public education. A natural science modeller teaches a select group about the scientific modelling of a local environmental problem. However, as we have also seen, understanding scientific modelling a bit more enables the local publics to communicate their points of view to decision makers. If they are successful in this, the participation could be seen as expressing a substantive rationale in the sense discussed in Chapter 3. The participation of lay people would have had an impact on what was decided.

The Community Modelling example highlights the complexity of environmental participation in the science-based expertise informing decision making. The binary of uninvited and invited participation was only the starting point. It captures the relationship between publics and expert-dominated environmental decision making along the axis of conflict or co-optation. It does not cover the development of independent expertise reflecting the priorities of environmental protection as in the case of American Rivers or other organisations that comprise local volunteers and scientific experts in a way that differs from both corporate experts and institutional decision makers. Nor does it capture the kind of environmental participation in expertise and decision making exemplified by the community modelling. In this practice increased understanding of scientific modelling tools contributed to empowerment in relation to local decision makers and opened up for more deliberative decision making.

This chapter has emphasised successful environmental participation in expertise informing decision making, from the Louisiana Bucket Brigade to Community Modelling in Otley. While it is important to highlight the ways in which environmental participation can lead to positive outcomes for local publics and environments there are also cause for caution. Expertise itself poses a challenge in democratic systems. There is a lively academic debate about how to handle environmental decision making underpinned by science-based expertise. While the issue may look unproblematic at first—expertise turns science into concrete advice on how to solve environmental problems—the gap identified by Nowotny between scientific knowledge and local circumstances needs thinking about. Which other knowledge and values are involved in the application of scientific principles to a local problem? The EJ movement in the United States draws attention to how corporate interests shape expertise in ways that disadvantages poorer communities. Local uninvited counter-expertise can give voice to some communities but the problem with the privileging of science-based knowledge claims in decision making is not solved.

At the other end of the scale, invited participation, CM and similar approaches have the potential to engage local people who have not previously been involved with environmental decision making. This is positive in the sense of increasing environmental participation and empowering local communities. Local environmental expertise can level the playing field which is easily dominated by institutional and corporate stakeholder interests due to their access to science-based expertise. However, when local communities become confident movers and shakers there is a risk that other stakeholders withdraw and leave it to local volunteer groups to take on responsibilities that they are unable to fulfil in the long run. Corporate and institutional actors may find it advantageous to fund local groups to do things like build wetlands at a fraction of the cost that full-time contractors would charge. That local residents get a chance to involve in volunteer activities is beneficial to all, but if this becomes a way for corporate actors to save money and avoid responsibility it is not a positive change.

Another challenge with the piecemeal redistribution of expertise that projects of the type that have been described in this chapter can offer is posed by their position in the democratic system. If formal democratic structures are bypassed by alliances of local stewardship groups and science-based expertise the decision making power could be diffused in a way that makes it very difficult to hold anybody accountable if something goes wrong. Not all expert advice lead to successful interventions and if failure occurs the traditional format of contracting technical expert consultants does have formal compensation measures available in the form of insurance and legally binding undertakings. The one-off interventions by environmental participation projects operate outside of this system. In addition, there are tensions between different scales where local engagement in one community could impact decision making in ways that have unintended negative consequences in another place (upstream and downstream is a common example). There can also be questions about effectiveness if local groups engaging via expertise focus exclusively on local interventions that will have very little impact on a larger scale problem, e.g. SUDS (Sustainable Urban Drainage Systems) will not suffice to prevent flooding in London if the Thames barrier is allowed to deteriorate.

The problems associated with the role of expert advice in democratic decision making take on new dimensions when expertise is redistributed by environmental participation. This is not a reason to refrain from such participation, but encouragement to think things through. Particularly involved scientists need to think much more about these aspects. In my experience local people are often aware of the challenges and limitations of local environmental participation in expertise. It is, after all, in relation to science-based expertise that they are understood as lay people, not with regard to environmental policy or local decision making. For scientists getting involved in environmental participation to redistribute expertise it is critical to learn about the social context from the local participants, to understand the possible consequences of the activities.

References

Jamison, Andrew. 2006. Social movements and science: Cultural appropriations of cognitive praxis. *Science as Culture* 15 (1): 45–59.

Landström, Catharina, Matilda Becker, Nicholas Odoni, and Sarah J. Whatmore. 2019. Community modelling: A technique for enhancing local capacity to engage with flood risk management. *Environmental Science & Policy* 92: 255–261.

Landström, Catharina, Sarah J. Whatmore, and Stuart N. Lane. 2011. Virtual engineering: Computer simulation modelling for flood risk management in England. *Science Studies* 24 (2): 3–22.

Nowotny, Helga. 2000. Transgressive competence: The narrative of expertise. *European Journal of Social Theory* 3: 5–21.

Ottinger, Gwen. 2010. Buckets of resistance: Standards and the effectiveness of citizen science. *Science, Technology and Human Values* 35 (2): 244–270.

Quet, Mathieu. 2014. Science to the people! (and experimental politics): Searching for the roots of participatory discourse in science and technology in the 1970s in France. *Public Understanding of Science* 23 (6): 628–645.

Whatmore, Sarah J., and Catharina Landström. 2010. Manning's N: Putting roughness to work. In *How well do facts travel? The dissemination of reliable knowledge*, ed. P. Howlett and M. Morgan, 111–135. Cambridge: Cambridge University Press.

Wilderman, Candie C., and Jinnieth Monismith. 2016. Monitoring Marcellus: A case study of a collaborative volunteer monitoring project to document the impact of unconventional shale gas extraction on small streams. *Citizen Science: Theory and Practice* 1 (1): 1–17.

Wynne, Brian. 2007. Public participation in science and technology: Performing and obscuring a political–conceptual category mistake. *East Asian Science, Technology and Society: An International Journal* 1: 99–110.

CHAPTER 5

Conclusion: Thinking Through Environmental Participation in Practice

Abstract In this concluding chapter, we start with reflecting on the preceding chapters and consider what the concepts used to organise the examples of environmental participation practices have brought to light. Then we revisit the social science thinking about participating publics to discuss how the examples prompt re-evaluation of the conceptual frameworks. After this we pick up on a thread introduced in Chapter 3 in the context of environmental participation in decision making—technologies of elicitation. This notion facilitates reflection of experts on participation and highlights resistance to public participation among some technical experts and decision makers. Finally, we consider the ways in which environmental participation relates to place.

Keywords Social science · Technologies of elicitation · Resistance · Place

Summing up the Examples of Environmental Participation Practices

This book was prompted by my perception that there was a gap in the academic discussions about public participation—no one seemed to address environmental participation as a particular practice. My STS colleagues do not distinguish public participation in environmental science and decision

© The Author(s) 2020
C. Landström, *Environmental Participation*,
https://doi.org/10.1007/978-3-030-33043-9_5

making from participation in other fields, such as health, or science and technology policy. Having done research in several transdisciplinary environmental projects, involving public participation for more than a decade, it seemed to me that environmental participation deserves more focussed attention than what has been the case so far.

As noted in Chapter 1 environmental participation mobilises diverse knowledges and it involves more-than-human agency. The environmental science and decision making in which publics participate are often prompted in response to non-human processes in nature, ranging from water pollution to biodiversity loss. Human actions may be the cause of environmental problems, but it is the ways in which nature reacts that require activities such as risk management. Environmental participation is also affected by the fact that scientists and experts do not have a monopoly on knowledge about environmental problems, people experience them in the everyday. In many cases environmental change, and the problems and risks that it brings, is visible to all. People can disagree with scientific knowledge claims about the environment based on their own experiences, observations and interactions with nature. In addition, environmental problems are time and place specific they occur somewhere at a particular time, in geographical locations where people live. These people may agree with the science-expert-decision making triumvirate on the best course of action, or they may not. Some of these people will have experience-based and practical knowledge about the local problem that surpasses that of any scientist or expert. Regardless of their knowledge people affected by an environmental problem have formal rights to participate in the decision making addressing it, in democratic countries. And, many environmental problems require ongoing monitoring and adaptive management which partly relies on local publics' capacity to contribute to the long-term governance. In my view, the particular combination of these features characterises environmental participation and sets it apart from public participation in other issues. The features as such can be found elsewhere, but their configuration is unique.

Within the field of environmental participation there is widely varying practices. Drawing on my own experience I made a pragmatic distinction between environmental participation in science, decision making and expertise. Experience has taught me that although scientists want their projects to inform decision making this is often not the case, which can leave involved lay participants very disappointed. While STS as a field understands science and society as entwined this does not mean that every scientific project impacts on decision making, or that all decisions about how to address environmental problems have a clear connection to academic science. In much of the decision making addressing environmental issues science features as expertise and intervening in the production of expert advice is one way for publics to participate. One source of differentiation of the field of environmental participation is the distinctive practices associated with science, decision making and expertise. Further variety exists within each domain. This book has brought together a wide range of activities not usually addressed within the same pages. However, differently from many academic discussions of public participation the focus here has been on practice. The environmental projects that I have experience of—doing participatory research and studying participatory decision making—made me think it would be useful to consider different practices as variation within three areas of environmental participation.

Several of the practices illuminated in this book come from projects with which I have personal experience. This has shaped this book from the outset, firstly I realised that my understanding of the historical background of environmental participation was very patchy, far from what historians would accept as reliable. Wanting to be upfront about this I retold some of the origin stories that I have come upon when involving with environmental participation. Interestingly these origin stories connected with particular environmental participation practices, as we found out in the three chapters on participation in science, decision making and expertise.

Learning from Public Participation in Environmental Science

In Chapter 1 the notions of 'co-production', 'dialogue' and 'education' (Callon 1999) commonly used in STS investigations, were understood as characterisations of the impact public participation is allowed to have on the scientific research in a project. Viewed in this way co-production is considered to be the most radical, involving scientists and lay participants collaborating as equals to create new knowledge. A distinctive feature of co-production was that it involves lay participants in the research activities in ways that impact on research questions, findings, analyses and outcomes.

To illustrate the notion of environmental participation as co-production we looked in detail at three environmental competency groups addressing water issues in the UK. This showed that although the format of all three was the same—scientists from social and natural science disciplines met to work with local participants at six scheduled meetings over one year—the groups developed differently. In all cases the initial plan had to be modified in some way. In the first competency group in Pickering the success of the new computer model settled the scientific outcome in the third meeting, from then on it became a question of detailing the new approach. In Uckfield the idea of a set group membership did not work and the group was opened up to new participants throughout all six meetings. In Marlborough the group generated findings that were fed into the wider process of developing a Neighbourhood Plan, rather than being turned into a freestanding output.

The second notion, dialogue, was illustrated with examples of stakeholder deliberation in one national UK project and one international EU project. In both projects stakeholder deliberation was constructed in a way that brought scientists and participants together separately from the research process to talk about their views on the issue at hand. The scientists presented their research and the stakeholders considered its relevance for their organisations and the problems they wanted to solve at specially arranged events, including seminars and workshops. The deliberation was not intended to change the science, but to create forums in which scientists and stakeholders could align their knowledge and their expectations. The outcomes of dialogue were exchange of data and scientific knowledge that would be useful to the involved stakeholders.

The third mode of interaction between science and publics was identified as education, however, further elaboration was needed to make the notion applicable to environmental participation in science. I argued that in the context of environmental participation this notion of education could be used with reference to citizen science. This is a form of participation in which the identity of participants does not matter, the involved publics are not expected to contribute knowledge in face-to-face interaction with scientists. On the contrary, some citizen science projects are fully mediated, with instructions of how to participate and what to do made available online. Anybody who is interested can participate in citizen science, lay people who want to join a project can arrive without knowledge and expect to learn when they get involved in the activity.

Applying Callon's (1999) three notions—co-production, dialogue and education—to the environmental participation practices I had encountered showed, as expected, that the most dramatic with regard to changing the way scientists worked was the co-production of knowledge. Involving lay people in the research process prompted many changes in how to conduct

research, for example, the local participants in the environmental competency groups constantly challenged the scientists to go out and look
for themselves rather than trust instrument data or published reports. The
stakeholder deliberation influence on the research was less radical to a great
extent because they were arranged to accommodate much larger collaborations, more scientists and more participants. There are limits to how
many people can work as close together as co-production requires, dialogue allows for many more people interacting. Even larger numbers can
take part in citizen science projects and they are not necessarily visible to
many of the researchers in projects relying on them. Had I not had a particular interest in how knowledge was generated in the drought project
I would probably not have noticed that the project relied on the citizen
science of Weather@Home.

One thing that the practical examples made clear was the fuzziness of the
boundaries between the three theoretical concepts used as an organising
framework. The threefold division of the relationship of science and publics
according to the models of co-production, dialogue or education worked
well in the beginning. They were useful to group environmental participation in science in three clusters, where the lay participants have different
roles. In the co-production example of environmental competency groups
lay participants took part in research activities, shaped research questions
and processes and produced outputs together with the scientists. At least
this is what they did some of the time, at other times the local participants
learned about how scientists calculate flood risk and how institutions do
cost-benefit analysis of flood prevention measures. On closer inspection
activities designated co-production also involved other modes of interaction, but not necessarily in the form of scientists acting in unison in relation
to a united publics. There was also education of the scientists regarding how
water moved through the particular places during flood and how land managers viewed different types of interventions. Occasionally subsets of the
transdisciplinary research team deliberated with institutional stakeholders
in the locality.

The stakeholder deliberation that I characterised as dialogue was also less
clearly delineated in practice. An interesting conceptual question emerged
regarding to what degree exchange of data involves co-production. In the
drought project stakeholders held important, proprietary, data about the
amount of water available at any time. Sharing this data with the research
project enabled the modellers to do things that they would not otherwise
have been able to do. In a reciprocal movement, the data generated by

the project was made into a publicly accessible resource at the end. This exchange seems much more like co-production than dialogue with regard to the impact on the research.

Not surprisingly citizen science is not simply a tool for education. Weather@Home generated data that was the driver for all the modelling in the drought project. The climate event sets that all modellers in the project were used to ensure that they worked with the same environmental phenomenon was generated through regional modelling by this citizen science initiative. Equally important on the scale of multiple research projects and national policy was the example of the Swedish Artportalen (The species portal).

Artportalen is built on the observations recorded by lay people, interested citizens with very varying degrees of specialist knowledge, ranging from avid birdwatchers to dogwalkers with a mobile phone app. Curated by scientists this data base has become the source material for scientific research in ecology and adjacent fields, it is also used to develop regulation and policy. I argued that both Weather@Home and Artportalen have become established as scientific infrastructure. Due to the ability of environmental research projects to handle large datasets and the increased availability of digitised environmental data the production of useful data has changed character in the last few decades making citizen science important in new ways.

Juxtaposing the environmental participation practices of co-production, stakeholder deliberation and citizen science the thought that comes to mind is that, perhaps the resource intensive, time demanding co-production projects impact more on the politics of the science–society relationship and the ways in which environmental scientists and local people relate than on scientific knowledge as such. In contrast, the mediated relationship of scientists and participants in large-scale citizen science, that invites everyone to take part, is already having a significant impact on the knowledge generation in environmental science.

Considering Public Participation in Environmental Decision Making

Turning to environmental participation in decision making the STS discussion provided another threefold conceptualisation. This one distinguishing three different rationales underpinning the invitation of publics to participate—normative, instrumental or substantive (Stirling 2007).

The normative rationale was understood to express a desire for a more just, democratic and equal system to govern environmental processes and risks. The big challenge for this rationale is to turn the ideals into practice and enable local publics to participate in meaningful ways. We traced the normative rationale in national and international environmental policy and noted that it was explicit in, for example the EU Water Framework Directive (WFD). According to the normative rationale public participation is a good in itself, as such it improves democracy and public influence and it does not need further justification.

The instrumental rationale is the opposite of the normative. It is a logic that encourages public participation because it is believed to facilitate the implementation of policy and institutional strategies. This rationale promotes participation in order to confer legitimacy on decisions already made, upon that the implementation will not be impeded by public opposition. A challenge for this rationale can be that it triggers public opposition as it is perceived as disingenuous, a different way of enforcing traditional top-down decision making.

The third—substantive rationale—insists that outcomes get better when more diverse ideas, knowledge and opinions are included in the decision making process. Invitations extended to publics premised on this rationale expect them to contribute new ideas for how to solve problems and fresh input to decisions about how to manage environmental challenges. The purpose is to access a wider range of knowledge and values. Practical and experience-based knowledge are both considered necessary for society to solve complex problems. One of many challenges to turning this into practice is the willingness of people to participate and their trust in the promises made about the openness to lay participants' ideas.

These three concepts were used in the discussion of the environmental participation practices of the West Cumbria MRWS (Managing Radioactive Waste Safely) Partnership and the Catchment Partnerships (CP) created within the Catchment Based Approach (CaBA) to involve local communities in river management in the UK. Upon closer examination of the examples we found that the different rationales were primarily associated with different actors and that rationales underpinning Government initiatives changed when they were put into practice by local actors in specific places.

With regard to the West Cumbria MRWS Partnership example the UK Government acted with an instrumental rationale, they saw participation as a way to succeed with siting a facility for geological disposal for radioactive waste. Previous attempts had been blocked by public opposition. The government thought that involving local communities from the start would make it possible to implement the strategy they had settled on decades ago. In contrast, the three local authorities that created the Partnership expressed a substantive rationale, they wanted accurate knowledge about the technoscientific dimensions and the community views on siting a geological disposal facility in the locality. The Partnership itself explicitly adhered to a normative rationale with an important objective being to make sure that every local resident got the opportunity to express their views.

The CaBA shows a similarly complex pattern, again the UK government was clearly acting with an instrumental rationale. Top-down government led by the Environment Agency (EA) and the water utility industry since the 1990s had obviously not succeeded in getting English rivers up to EU WFD water quality standards. To improve water quality and implement the WFDs explicit requirement of local community involvement the Government created a new environmental entity, the Catchment Partnership. The actors coming together in CPs all want to effectuate catchment management that will improve water quality and, although they may not necessarily agree on what the best knowledge and techniques are, this expresses a substantive rationale. The civil society groups acting as CP hosts are also committed to engage local residents with rivers and the natural environment when plans are made for managing a catchment, while institutional actors such as the EA could consider the groups themselves to satisfy the need for public representation.

That the three rationales blend in different ways in relation to particular practical examples was interpreted as showing the complexity of governance. If there was an easy way to implement top-down environmental policy public participation in the decision making on geological disposal and catchment management would be purely instrumental. The government would be seeking legitimacy and consensus for its decisions. However, environmental decision making is not top-down anymore, instead we are confronted with multilevel, multisited governance with a variety of actors, all impacting on environmental decision making.

The STS discussion has, so far, not discussed environmental participation as an issue that reconfigures aspects of environmental governance. In this book, I have indicated that this is a topic worth considering as we have seen that participation can become a way to bring together the institutional stakeholders and local communities involved in decision making in new formal collaborations. The two examples of public participation in environmental decision making in this book show that there is more to decision making than Government ambition. The distributed and complex decision making processes in environmental governance require analyses that move beyond the emphasis on institutional intent seemingly intrinsic to the concept of the three rationales. The conceptual toolbox needs further development to capture the dynamics of public participation in environmental governance.

Complex Entanglements of Environmental Participation and Expertise

The complexity of the relationship of science and decision making came into focus in Chapter 4 when the notions of uninvited and invited participation (Wynne 2007) were used as a starting point for looking at environmental participation in expertise. This was informed by a view on environmental participation as a way for publics to challenge the science-based expertise that often underpins decision making.

Uninvited challenges to knowledge claims are rarely experienced by environmental scientists because their research usually does not lead to concrete suggestions for local interventions. Scientific field studies move from data collected in local environments to general theory. Environmental scientists can study, for example, a river, with the intent of creating a computer simulation model that can be used to analyse water quality in all rivers with similar features. Scientific knowledge strives for generality, local conditions are treated as instances of generally occurring processes. In contrast, technical experts apply generalised scientific knowledge to local problems. A technical consultant would use a computer model created by scientists to analyse the water in a particular river and then produce a report that answers the questions the commissioning actor are interested in. If the actors giving the experts their brief disregard the local community and push ahead with activities that people would consider as damaging they could challenge the expertise underpinning the decision. Challenging expertise can be an effective route for publics to influence decision making.

The notion of uninvited and invited participation highlights the politics of knowledge. Its original formulation does not distinguish between science, decision making and expertise in the way this book does. Uninvited participation appears to refer to publics challenging the establishment (which is a merger of science, societal institutions and corporate interests). This interpretation seems reasonable given the definition of invited participation as always subjected to imposed framing of the issue in ways that serve the powers that be. However, it is not necessary to adopt the world view undergirding a concept in order to use it and we used the notions of uninvited and invited participation begin the examination of environmental participation and expertise.

First we looked at a well-known example of uninvited participation—the Louisiana Bucket Brigade—an example showing how a local community's concern about air quality was initially dismissed by the polluting industry and their experts. But, when getting assistance from critical scientists, local activists were able to produce counter-expertise and effectively challenge the knowledge claims and the decision making. This example is one of the success stories of the environmental justice movement that we understand as shaped by the particular constellation of science, law and capital in the United States.

Bringing the binary of uninvited and invited participation into question the second example in Chapter 4 was of another US organisation: American Rivers. This and similar non-profit, volunteer-based organisations existed in the environmental domain when environmental science was a new field. Some of these organisations have pursued science-based investigation of local problems since the early days and they have developed expertise with the aim of protecting the environment. These organisations are rarely mentioned in STS discussions of public participation, but must be understood as important sites for environmental participation.

A mirror image of environmental movement expertise was provided in the example of academics initiating citizen science as local action to monitor local environmental impacts of shale gas extraction in the ALLARM project. This example troubles the idea of scientists as neutral in relation to decision making, as well as another common assumption that scientists will always side with industry. The focus on US examples of uninvited, movement-based and scientific activism came about because of the relative clarity of this adversarial system. Challenging institutional and corporate activities in court makes the role of expertise clearly delineated. In many other countries the ways in which expertise relates to decision making is less explicit, making case studies longer and more complicated to draw conclusions from.

Circling back to the binary of uninvited and invited environmental participation we finished Chapter 4 by looking at Community Modelling as a practice that can help local communities to use scientific tools to strengthen their capacity to participate. Originating in the experience of environmental competency groups as a participatory method in science Community Modelling uses computer models to improve the capacity of local communities to take part in environmental decision making by learning about the tools used by technical experts. Embracing four principles—resource minimalism, strategic recruitment, extant software and connecting with decision makers—Community Modelling attempts to make environmental science work for local environmental groups. In the complex webs of environmental governance institutions and corporate actors can often access scientific knowledge by commissioning technical experts to produce the knowledge they need. This is not possible for local environmental groups, and that can create barriers for them to accept invitations to participate and to be effective when participating in decision making.

Reflecting on the different ways in which publics can participate in environmental expertise we noted that the empowerment of local people in relation to the decision making authorities raises new challenges. Expertise does in itself undermine democratic decision making since the choices underpinned with it tend to count more. Balancing corporate expertise with counter-expertise, as is the case in environmental justice, is not a reliable path to democratic environmental management.

It is challenging to align environmental participation through expertise with formal decision making in representative democracy. Decisions about how to allocate funding and address problems are expected to be taken by elected representatives with well-defined mandates in society. Decision making bodies, such as district councils, comprise representatives elected by all citizens eligible to vote and choosing to do so. The elected representatives hold their posts for a set time and are expected to both voice the interests of their voters and consider the common good. However, if some environmental groups gain more influence by participating in the creation of expertise that underpins decision making this can lead to unfair distributions of environmental risks and benefits in a community.

We also note that environmental participation in expertise attracts the people who are interested and able to spend time and effort. People who do not engage with formal organisations rarely get involved with environmental decision making. Environmental participation in expertise tend to attract better educated, more affluent, older people. Hence, environmental participation could reinforce local inequality, marginalising the most marginalised even further. In a worst-case scenario environmental participation in expertise could lead to alignment of the goals of environmental groups and corporate actors, swaying decision making at the cost of non-participating residents.

The problems associated with expert advice in democratic decision making take on new dimensions when expertise is redistributed in environmental participation. This is not a reason to refrain from such participation, but encouragement to further investigation and reflection.

Participating Publics

In Chapter 1 we raised the question about the identity of participating publics. There are many discussions of this in the literature and conceptualisations abound. An idea that I find useful is Sally Eden's notion of knowing publics (Eden 2017). When publics are invited to co-production of knowledge in some form it must be assumed that they know things about the environment that scientists do not. There are several terms in use to denote the knowledge of publics, for example, lay- or local. In this book different terms have been used but what unites them is the idea that this knowledge is based in people's experience, not as scientific knowledge in systematic investigation or experiments.

There are many different ways to learn about the environment, according to Eden 'the outdoor passions of rock-climbers, anglers, birdwatchers and gardeners will vary both within and between each group, as well as differ from the indoor environmental pursuits of those watching nature documentaries on television' (Eden 2017: 36). We could add to this list participate in scientific projects, decision making or creation of expertise. Environmental participation is a way for, primarily already interested people, to become particular knowing publics. Still, everybody brings their prior knowledge and experience to environmental participation in which this is more or less important. For example, in contrast to co-production citizen science does not require any pre-existing knowledge among participants.

Co-production of knowledge and citizen science involves publics with background knowledge different from that of the scientists, in contrast stakeholder deliberation can address professionals with similar educations but active in different fields. I have used the notion of stakeholder for a particular type of publics in particular, to use Mike Michael's term. Stakeholders are organised and 'have an identifiable stake in particular scientific or technological issues or controversies' (Michael 2009: 623). Stakeholders were understood to be identified through pre-existing, independent relationships with the environmental issue at hand. Publics in particular can be of many different types, local environmental groups, private businesses, tenants' organisations, homeowners, local authorities, recreational groups and so on, importantly it is by being organised that they become visible to scientists and decision makers as stakeholders.

Stakeholders feature in environmental participation in both science and decision making. In both fields, they are expected to have their own interests and agendas that make them willing to participate and to care about the outcomes of a scientific project or a decision making process. Stakeholder organisations are often represented by professionals or in-house experts when participating in science or decision making. Hence, stakeholder deliberation can in some cases result in closed circuits of expertise that exclude other publics from participating.

The practices discussed in the previous chapters brought to light different relationships between actors positioned as stakeholders in varying contexts. In relation to scientific research experts could be involved as representatives of organisations to deliberative events. As well as voicing their organisation's interests they could supply information contributing to the generation of new scientific knowledge. In decision making their role would be different, formal stakeholders represented by professionals were closely associated with the decision making bodies. In the examples discussed in Chapter 3 the stakeholders were institutions and corporate actors pursuing their own agendas. In Chapter 4 we saw the experts acting as independent advisors, translating scientific knowledge to actionable proposals.

In many environmental decision making processes experts play a key role. Technical experts provide knowledge that decision makers use in processes that often exclude civil society. This is the kind of situation in which civil society can organise and engage in uninvited participation. In this chapter we looked first at publics involved in environmental justice in the United States. Often these groups arise in socioeconomically disadvantaged communities that are disproportionally subjected to environmental risks. They engage in environmental participation for reasons other than an interest in the natural environment. In contrast, organisations like American Rivers are all about volunteering as a way of caring for the natural environment. Such groups pursue their own agendas with regard to science and decision making and they systematically promote expertise that is not governed by industrial or economic growth interests. In the example of Community Modelling we saw how a participatory technique could prompt public involvement by supporting local volunteer organisations' ability to engage with decision making.

Overall the practices discussed in this book demonstrate the diversity of environmental publics that participate in science and decision making. The character and identity of the participating publics are to a significant extent shaped by the specific technology of elicitation—the position they are made to fill in a particular practical organisation of participation. Participation formats, as such, have become subjects of expertise.

Professionalisation of Environmental Participation

A new development in the field of environmental participation is the advent of experts on participation.[1] These experts are primarily social scientists who have specialised in the field of public participation and developed effective formats that are used to organise participation events. While not exclusive to environmental participation the policies and regulations requiring public involvement with environmental governance provide a lot of work for these experts.

In the previous we saw an example of this in Chapter 2, where the West Cumbria MRWS Partnership commissioned several expert consultants to organise face-to-face events in the local community. 3QK, a well-established consultancy had a particularly prominent role, working with the Partnership on all aspects of the programme. The Partnership members were lay people in relation to eliciting community views, as well as with regard to natural science on radioactive waste. They recognised the need for expertise in order to achieve their objectives of comprehensively establishing the community's view on siting a GDF in the locality. While they could rely upon the formal stakeholders and the government agencies involved in the GDF siting process to provide natural science and engineering expertise they had to turn elsewhere for expertise on public engagement. In contrast public participation in science is often a research undertaking in itself. In interdisciplinary environmental projects it has become common to have a social science team doing research on participation as a social aspect of the problem. These research projects can involve Ph.D. students and postdocs who afterwards go on to become experts and perhaps work as consultants.

Professional participation experts contribute their skills to make public participation a regular feature of environmental decision making. Whether the events they organise express normative, substantive or instrumental

---

[1] A recent anthology edited by Bherer et al. (2017) examines this phenomenon.

rationales depends on the context. The participation consultants working for the Partnership were commissioned to organise events that would allow the public to impact on the decision making. In other cases the motive may be different, particularly if inviting the public was not really a choice. In these cases commissioning participation expertise become a way to demonstrate compliance without letting publics affect the outcomes of the decision making process.

Resistance to Environmental Participation

Studies of public participation in science and technology policy and environmental governance have noted how the formats of participation can constrain the ability of the participating publics to impact on the outcome of the processes. To control and contain participation would be in the interest of decision makers who are compelled by policy and regulation to facilitate it, but who do not consider it useful.

Case studies of environmental managers have found that many are considerably less positive to public participation than the policy frameworks governing their activities. Wesselink et al. (2011) identified a fourth rationale for participation that they called 'legalistic' and according to which 'participation is only organised to meet formal requirements' (Wesselink et al. 2011: 2691). Participation organised with this intent 'is likely to be a formality without any uptake of results' (Wesselink et al. 2011: 2691). It is reasonable to assume that many instances of mandatory public consultation on environmental decision making are informed by a legalistic rationale.

To deflect the potential influence of public participation is not unique to environmental participation. Kathrin Braun and Susanne Schultz (2010) discuss how the construction of participating publics can contradict the stated purpose of contributing to decision making. Studying participatory governance arrangements in the area of genetic testing they found that 'the main purpose of the participatory processes we looked at are knowledge production and education rather than political deliberation and decision-making' (Braun and Schultz 2010: 415). To substitute participation in decision making with education of the public can be a way to mollify opposition. However, publics are not necessarily compliant.

Invited publics can resist attempts to keep them contained. Ulrike Felt and Maximilian Fochler (2010) highlights the way in which participating publics can appropriate and transform the roles and identities they are assigned. They make a finding that is quite significant for the present book noting that 'the participating citizens drew on shared implicit resources reflecting the techno-political culture they are part of' (Felt and Fochler 2010: 235). The people participating in environmental science, decision making and expertise are not devoid of knowledge. They may not be environmental scientists, decision makers or experts, but they often have other scientific, professional or political knowledge and skills that they can draw upon to defend their right to be heard. That the participating lay people are not passive blank slates is an important point and that their expectations are shaped by the wider culture is also significant. In the context of the examples of environmental participation discussed in this book it provides us with ideas about how to understand public interest in participating and what they might expect to gain.

The Place of Environmental Participation

When beginning to consider not only the ways in which participating publics are produced but also why people step forward to join such a public we return to the starting point of this book—that environmental participation is particular. In a very concrete sense there is no such thing as a general public in environmental participation. Every individual who participates has a relationship with the environment that prompts them to participate in science, decision making and/or expertise. In all the examples of environmental participation practices that have been discussed in the three middle chapters of this book, one thing is clear—interest in our relationship with the natural environment is what motivates individuals to take part. From the modelling water industry experts who want to be able to access new data in order to better understand how they can use the resource in a more sustainable way, to the pensioners volunteering to build new wetlands in the rivers of London, to members of local environmental groups attending Catchment Partnership meetings—they care about environmental matters.

The many people comprising these publics have very different relationships to the natural environment but, one important uniting feature is a relation to place. Experience-based environmental knowledge and concerns are primarily place-based. It is the environment in a particular place that one can experience, the abstractions featuring in, for example, climate change discourses, e.g. average global temperature, cannot be experienced anywhere. In contrast, flooding or pollution events occur in actual localities. Regardless of how environments are experienced in diverse practices (rambling, rowing, swimming, angling, gardening, protesting, etc.) it will happen in a particular place. This matters greatly for environmental participation. The connection of experiences with place in environmental participation is key to successful engagement. It is important to understand that in this field, lay knowledge is local. It is equally important to understand that scientific and expert knowledge is partial in other ways. Scientists environmental knowledge is limited by methodology and discipline, today many scientists experience the environment through computer modelling of systems in which the phenomena they are interested in can be represented. The physical environment is present as data sets retrieved from servers. Scientists working in this way rarely experience the limitations of their knowledge with regard to local problems since their mathematical constructs are general and their models can represent any circumstances. When scientists working in this way meet with publics in participatory research the two can fail to understand each others different relations to place.

Scientists are trained to value knowledge about general principles and mechanisms over knowledge about the unique and particular. In environmental sciences this is expressed in the understanding of research sites as providing data that can be used to construct general knowledge and models. The epistemic role of local environments in science is to provide information about examples of general principles. This approach aligns with environmental participation in the form of citizen science. This way of working provides scientists with data of a much larger number of local occurrences of phenomena than what they would otherwise have access to. Citizen science makes no demands on scientists to change.

Co-production can be very challenging for scientist who learn that their orientation towards the general becomes an impediment in collaboration with local people. In practice co-production requires that research objectives can be allowed to diverge from that which scientists would regard as important. This will entail a different valuation of knowledge about the unique and particular ways in which an environmental process unfolds in the locality. This is only one of the ways in which co-production requires that science is done differently and it does not guarantee successful participatory research, but it is a necessary condition.

Stakeholder deliberation is a type of environmental participation that we find in both science and decision making. Inviting publics-in-particular enables scientists and decision makers to collaborate with publics whose explicit interest in the knowledge or decision is easy to identify and it provides a focus and framework for participation. In this context, the abstract notion of scale is appropriate because the actors inviting to environmental participation are in a position to identify stakeholders who exercise agency across the same geographical area. However, in practice further delegation of tasks, such as in the case of the West Cumbria MRWS Partnership can bring back a focus on a unique place.

Place is firmly put in the centre by uninvited participation. In such cases, the particularity created by local circumstances is used to challenge the ways in which experts apply general scientific knowledge. Counter-expertise presents different knowledge claims about the place that establishment experts have claimed that they know. If such different knowledge claims about a place are created in collaboration with scientists such challenges have a very good chance of success. Thankfully, not all local participation in expertise has to be prompted by conflict, we also had examples of how independent environmental organisations could become knowledge providers. And, of how scientific and expert tools can be opened up to local participation in ways that support public participation in environmental decision making.

EPILOGUE

This book was motivated by my perception of a lack. I thought environmental participation qualified as a distinct field of practice. Beginning with thinking about the justification for this pushed my attention towards the multitude of participatory activities going on in environmental science and decision making. My perceptions were based on personal experience of working with local people and natural scientists in transdisciplinary research projects and doing STS case studies of environmental governance. It felt important to think this through as elements in a coherent narrative, hence, the notion of environmental participation came to encompass a broad range of activities. Ironically clear and distinct concepts are needed to organise stories about infinite variety and blurry boundaries. I turned to some of the concepts that are used most frequently by STS scholars discussing participation and environmental processes. These well-known concepts facilitated the presentation of diverse examples of environmental participation practices in ways that made sense. To distinguish between participation in environmental science, decision making and expertise was important to organise the discussions of practical activities and how public participation can impact on institutions. Interestingly putting these concepts to use in discussing different practices also brought to light their limitations.

Hopefully, the discussions in this book have managed both to present examples of environmental participation in practice in an illuminating way and demonstrated the possibilities and limitations of some commonly used social science concepts. I also hope that environmental participation has been clarified as a type of activity worth pursuing as research topics and as practices.

References

Bherer, Laurence, Mario Gauthier, and Louis Simard (eds.). 2017. *The professionalization of public participation*. New York: Routledge.

Braun, Kathrin, and Susanne Schultz. 2010. '... a certain amount of engineering involved': Constructing the public in participatory governance arrangements. *Public Understanding of Science* 19 (4): 403–419.

Callon, Michel. 1999. The role of lay people in the production and dissemination of scientific knowledge. *Science Technology & Society* 4 (1): 81–94.

Eden, Sally. 2017. *Environmental publics*. London and New York: Routledge.

Felt, Ulrike, and Maximilian Fochler. 2010. Machineries for making publics: Inscribing and de-scribing publics in public engagement. *Minerva* 48 (3): 219–238.

Michael, Mike. 2009. Publics performing publics: Of PiGs, PiPs and politics. *Public Understanding of Science* 18 (5): 617–631.

Stirling, Andy. 2007. Opening up or closing down? Analysis, participation and power in the social appraisal of technology. In *Science and citizens. Globalization and the challenge of engagement*, ed. M. Leach, I. Scoones, and B. Wynne, 218–231. London and New York: Zed Books.

Wesselink, Anna, Jouni Paavola, Oliver Fritsch, and Ortwin Renn. 2011. Rationales for public participation in environmental policy and governance: practitioners' perspectives. *Environment and Planning A: Economy and Space* 43 (11): 2688–2704.

Wynne, Brian. 2007. Public participation in science and technology: Performing and obscuring a political-conceptual category mistake. *East Asian Science, Technology and Society: An International Journal* 1: 99–110.

INDEX